Advanced Fuzzy Systems
Design and Applications

Studies in Fuzziness and Soft Computing

Editor-in-chief

Prof. Janusz Kacprzyk
Systems Research Institute
Polish Academy of Sciences
ul. Newelska 6
01-447 Warsaw, Poland
E-mail: kacprzyk@ibspan.waw.pl
http://www.springer.de/cgi-bin/search_book.pl?series=2941

Yaochu Jin

Advanced Fuzzy Systems
Design and Applications

With 170 Figures
and 10 Tables

Physica-Verlag

A Springer-Verlag Company

Dr. Yaochu Jin
Honda R&D Europe GmbH
Future Technology Research
Carl-Legien-Straße 30
63073 Offenbach/Main
Germany
yaochu.jin@hre-ftr.f.rd.honda.co.jp

ISSN 1434-9922
ISBN 978-3-7908-2520-6

Library of Congress Cataloging-in-Publication Data applied for
Die Deutsche Bibliothek – CIP-Einheitsaufnahme
Jin, Yaochu: Advanced fuzzy systems design and applications: with 10 tables / Yaochu Jin. – Heidelberg;
New York: Physica-Verl., 2003
    (Studies in fuzziness and soft computing; Vol. 112)
    ISBN 978-3-7908-2520-6   ISBN 978-3-7908-1771-3 (eBook)
    DOI 10.1007/978-3-7908-1771-3

Physica-Verlag Heidelberg New York
a member of BertelsmannSpringer Science+Business Media GmbH

© Physica-Verlag Heidelberg 2003
Softcover reprint of the hardcover 1st edition 2003

SPIN 10892920   88/2202-5 4 3 2 1 0 – Printed on acid-free paper

Preface

Fuzzy rule systems have found a wide range of applications in many fields of science and technology. Traditionally, fuzzy rules are generated from human expert knowledge or human heuristics for relatively simple systems. In the last few years, data-driven fuzzy rule generation has been very active. Compared to heuristic fuzzy rules, fuzzy rules generated from data are able to extract more profound knowledge for more complex systems.

This book presents a number of approaches to the generation of fuzzy rules from data, ranging from the direct fuzzy inference based to neural networks and evolutionary algorithms based fuzzy rule generation. Besides the approximation accuracy, special attention has been paid to the interpretability of the extracted fuzzy rules. In other words, the fuzzy rules generated from data are supposed to be as comprehensible to human beings as those generated from human heuristics. To this end, many aspects of interpretability of fuzzy systems have been discussed, which must be taken into account in the data-driven fuzzy rule generation. In this way, fuzzy rules generated from data are intelligible to human users and therefore, knowledge about unknown systems can be extracted.

The other direction of knowledge extraction from data in terms of interpretable fuzzy rules is the incorporation of human knowledge into learning and evolutionary systems with the help of fuzzy logic. In this book, methods for embedding human knowledge, which can be represented either by fuzzy rules or fuzzy preference models, into neural network learning and evolutionary multiobjective optimization have been introduced. Thus, neural networks and evolutionary algorithms are able to take advantage of data as well as human knowledge.

In this book, fuzzy rules are designed mainly for modeling, control and optimization. Along with the discussion of the methods, several real-world application examples in the above fields, including robotics, process control and intelligent vehicle systems are described. Illustrative figures are also given to accompany the most important methods and concepts. To make the book self-contained, fundamental theories as well as a few selected advanced topics about fuzzy systems, neural networks and evolutionary algorithms have been provided. Therefore, this book is a valuable reference for researchers, practitioners and students in many fields of science and engineering.

A major part of the book is written during spare time. Therefore, I would like to thank my wife, Fanhong and my two little children, Robert and Zewei for their understanding and encouragement. I am very thankful to Dr. Bernhard Sendhoff for inspiring discussions on several topics in the book during my work at Ruhr-University Bochum and at Honda R&D. Finally, I am grateful to Prof. Werner von Seelen, Prof. Edgar Körner and Prof. Jingping Jiang for their kind support.

Offenbach am Main, June 2002   *Yaochu Jin*

Contents

1. Fuzzy Sets and Fuzzy Systems

1.1 Basics of Fuzzy Sets

1.1.1 Fuzzy Sets

A set is a collection of finite or infinite number of objects, which are called elements or members of the set. In the classical set theory, an object either belongs to a set or does not. If x is an element of set A, we notate $x \in A$; if not, we notate $x \notin A$.

Generally, there are two methods for describing the characteristics of a particular set:

- The listing method. In this approach, we describe a set by listing its elements, for example, $A=\{1,3,5\}$ or $B=\{$bachelor, master, doctor $\}$.
- The membership rule method. In this approach, a set is described by a membership rule, which determines whether an object is a member of the set. For example, $A=\{x \mid x$ is an odd integer that is larger than 0 and smaller than 6 $\}$ or $B=\{y \mid y$ is an academic degree $\}$.

The membership rule method can be realized, among others, by a characteristic function. Suppose set A is a subset of universe U, $x \in U$, then whether x is a member of set A can be determined by the following function $\Phi_A(x)$:

$$\Phi_A(x) = \begin{cases} 1, & \text{if } x \in A, \\ 0, & \text{if } x \notin A. \end{cases} \qquad (1.1)$$

For example, let $A = \{x \mid x$ is a real number between -1 and $+1\}$, then the following characteristic function can be used to define set A:

$$\phi_A(x) = \begin{cases} 1, & \text{if } -1 \le x \le +1, \\ 0, & \text{else}, \end{cases} \qquad (1.2)$$

which is shown in Fig. 1.1.

From Fig. 1.1, we notice that for any real number, its membership for set A is uniquely determined by the characteristic function, either equals 1 or 0. Therefore, in the classical set theory, a definition of a concept (set) admits of no degrees.

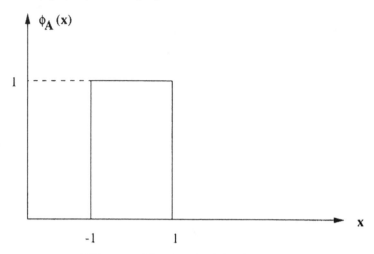

Fig. 1.1. The characteristic function.

However, this contradicts strongly with the real world. As a matter of fact, a lot of concepts, especially those used by human beings, possess a grade of degree. Concepts such as mountain, hill, lake, pond and so on are very hard to be described by a classical set. A lot of attributes, such as, low, high, short, tall, beautiful, ugly and so on, are ambiguous. If we intend to describe *tall men* with a classical set, we have to provide a precise standard, say, 180 cm or taller. Then the situation will be that a person whose height is 180.2 cm will be classified as tall men whereas another one who is 179.8 cm will be not. This seemingly precise way of classification brings about an unreasonable result.

Let us consider an another example. In driving, if one meets a red light at an intersection, one has to brake to stop the car. If the speed of the car is low and the car is still far from the intersection, one only needs to brake slightly. However, if the speed is very high and the car is very close to the intersection, one has to brake the car firmly. In this process of decision making, no precise data are involved in. It is absurd if one makes a decision based on the exact reading of car speed and the exact number of distance.

The classical set theory also encounters great difficulties in describing the relationship between two concepts. In Fig. 1.2, two acoustic sensors (denoted by 'X') are deployed in the concerned area, which is partitioned into a number of grids. Under normal weather conditions, the region that the sensor can detect is a circle with a radius R. Now, let $A=\{x \mid x$ is a sensor that can see grid $G\}$. By using the classical set theory, we are at loss to give an answer, because the classical definition of belonging is unable to describe the relationship between a square and a circle.

The limitations of the classical set theory lies in the fact that a characteristic function that describes a classical (crisp) set can only assume 0 or 1.

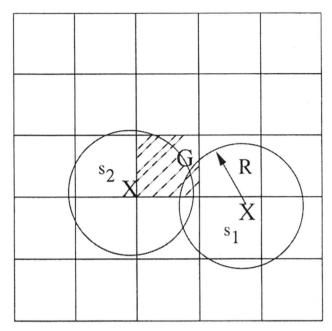

Fig. 1.2. Sensors that can see grid G.

However, if we allow it to take a value between 0 and 1, these difficulties can be removed. This pivotal idea was first proposed by Lotfi A. Zadeh in [253].

Let A be a set on the universe U described by the characteristic function $\mu_A(x)$:

$$\mu_A(x) : U \longrightarrow [0, 1] \qquad (1.3)$$

for any $x \in U$, $\mu_A(x) \in [0, 1]$ is a function that specifies a degree to which element x belongs to set A. The set A is called a fuzzy set and the characteristic function $\mu_A(x)$ is called a function.

According to the definition of a fuzzy set, an element belongs to a set with a grade of degree. The maximal membership is 1 and the minimal is 0. Therefore, we can see that a fuzzy set is an extension of a classical set, where a classical set is a special case of a fuzzy set.

Now we can define *tall men* with a fuzzy set. Fig. 1.3 provides an example. We notice that in this description, a person who is 1.70 *cm* tall belongs to the fuzzy set *tall men* with a degree of 0.60, whereas another one who is 1.80 *cm* tall belongs to *tall men* with a degree of 0.90. This is no doubt much more sound than we say that the latter is tall and the former is not.

We are now also able to answer the question whether a sensor can see a grid that we have discussed previously (refer to Fig. 1.2). For the two sensors s_1 and s_2, it is a reasonable answer to say that the sensor s_i can see the grid G to a degree of

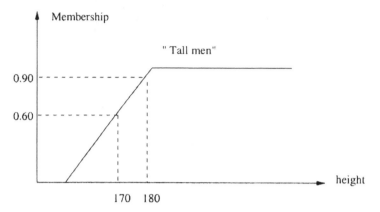

Fig. 1.3. A definition of fuzzy set "tall men".

$$d_i(s_i) = \frac{\text{Area}(s_i) \bigcap \text{Area}(G)}{\text{Area}(G)}, i = 1, 2, \qquad (1.4)$$

where, $Area(s_i)$ is the area the sensor s_i can see, $Area(G)$ is the area of the grid G, and '$\bigcap$' denotes the overlapping of the two areas. Thus, for the two sensors in Fig. 1.2, a reasonable answer is, approximately, that the sensor s_1 can see the grid G with a degree of 0.1, and the sensor s_2 can see the grid G with a degree of 0.8.

Mainly, there are three methods for representing a fuzzy set. Assume U is the universe which is a classical set with n elements:

$$U = \{u_1, u_2, ..., u_n\}, \qquad (1.5)$$

A is a fuzzy set on the universe U specified by a membership function $\mu_A(u)$, then fuzzy set A can be expressed in one of the following three ways:

- $A = \mu_A(u_1)/u_1 + \mu_A(u_2)/u_2 + ... + \mu_A(u_n)/u_n$. Note that symbol '/' is not a division and symbol '+' is not a plus.
- $A = \{(u_1, \mu_A(u_1)), (u_2, \mu_A(u_2)), ..., (u_n, \mu_A(u_n))\}$. It can be seen that a fuzzy set is described by a series of ordered pairs.
- $A = (\mu_A(u_1), \mu_A(u_2), ..., \mu_A(u_n))$. Note that in this notation, a membership degree can not be left out from the vector even if it is zero.

If universe U is continuous and cannot be described by a finite number of elements, the following notation is used to describe fuzzy set A:

$$A = \int_U \mu_A(u)/u. \qquad (1.6)$$

Similarly, symbol '$\int$' does not mean integration, but means a 'collection of'. To illustrate the previously defined notations of fuzzy sets, we present here a simple example. Let $U = \{0, 1, 2, 3, 4, 5, 6, 7, 8, 9, 10\}$, A is a fuzzy set:

$$A = \{i| \ i \text{ is close to } 5, i \in U\}. \tag{1.7}$$

If the fuzzy membership function is defined as in Fig. 1.4, then fuzzy set A can be written as:

$$A = 0/0 + 0.2/1 + 0.4/2 + 0.6/3 + 0.8/4 + 1/5$$
$$+0.8/6 + 0.6/7 + 0.4/8 + 0.2/9 + 0/10. \tag{1.8}$$

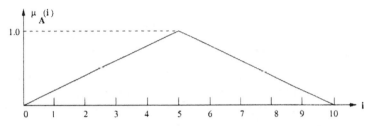

Fig. 1.4. A definition of the fuzzy set "close to five".

If the fuzzy set is expressed by ordered pairs, it looks like:

$$A = \{(0,0), (1,0.2), (2,0.4), (3,0.6), (4,0.8), (5,1.0),$$
$$(6,0.8), (7,0.6), (8,0.4), (9,0.2), (10,0)\}. \tag{1.9}$$

Finally, if A is notated by a vector, it has the following form:

$$A = (0, 0.2, 0.4, 0.6, 0.8, 1.0, 0.8, 0.6, 0.4, 0.2, 0). \tag{1.10}$$

The properties of a fuzzy set are uniquely determined by its membership function. Usually, the elements on the universe whose membership is larger than zero are called the *support* of the fuzzy set. If there is only one element in the support of the fuzzy set, then this fuzzy set is called a *fuzzy singleton*. For a fuzzy set A on the universe U defined by a membership function $\mu_A(x)$, $hgt(A)$ is called the *height* of the fuzzy set that is defined by:

$$\text{hgt}(A) = \max_{x \in U}\{\mu_A(x)\}. \tag{1.11}$$

If $hgt(A) = 1$, then the fuzzy set A is said to be *normal* and all its elements that have the a membership degree of 1 are called the *kernel* of the fuzzy set. A fuzzy set A is *non-normal*, if $0 < \text{hgt}(A) < 1$. Nevertheless, a non-normal fuzzy set can be normalized by dividing the membership function $\mu_A(x)$ by the maximal membership grade, i.e.,

$$\mu_A(x) = \frac{\mu_A(x)}{\text{hgt}(A(x))}. \tag{1.12}$$

A fuzzy set A is called an empty set $(\emptyset)$ if $\mu_A(x) = 0$ for all $x \in A$.

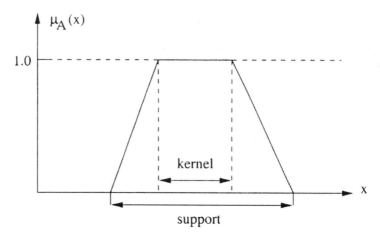

Fig. 1.5. Support and kernel of a fuzzy set.

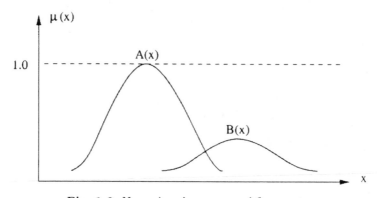

Fig. 1.6. Normal and non-normal fuzzy sets.

Fig. 1.5 illustrates the support and the kernel of a fuzzy set, Fig. 1.6 is an illustration of a normal fuzzy set A and a non-normal fuzzy set B.

Fuzzy sets are suitable for describing sets whose boundaries are not sharply defined. The development of fuzzy sets has opened up several new research fields in mathematics, system theory, logic and cognitive science. It provides an effective way of dealing with uncertainties other than the probability theory.

Because of the wide range of uncertainties in human knowledge, it is in some cases insufficient to use a precise, deterministic membership function. Therefore, one can also use *type-2* fuzzy sets, which were first introduced by Zadeh [254] as an extension to the conventional fuzzy set (also known as *type-1 fuzzy set*). For type-2 fuzzy sets, their membership is again a type-1 fuzzy set. A fuzzy set of type-2 A in a set U is the fuzzy set characterized by the fuzzy membership function μ_A as

$$\mu_A : U \longrightarrow [0,1]^{[0,1]}, \tag{1.13}$$

where $\mu_A(x)$ is known as a fuzzy grade of a fuzzy set in $[0,1]$. For example, A is a fuzzy set "close to five" on the universe of the discourse $U = \{0,1,2,3,4,5\}$. Instead of defining the membership degrees with crisp numbers, we define the membership degrees with fuzzy sets as follows:

$$A = \{\text{very far}/0 + \text{far}/1 + \text{not far}/2 + \text{quite close}/3 + \text{close}/4 + \text{very close}/5\}. \tag{1.14}$$

We notice that "very far", "far", "not far", "quite close", "close" and "very close" themselves are fuzzy sets.

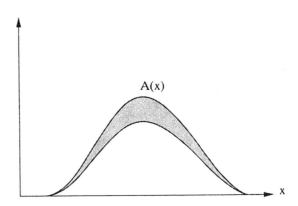

Fig. 1.7. An interval-valued fuzzy set.

Let us take a look at an another example of a type-2 fuzzy set that has been provided in [167]. Let set $U = \{$Susie, Betty, Helen, Ruth, Pat$\}$ and A is a fuzzy set of "beautiful women" on U:

$$A = \text{middle}/\text{Susie} + \text{not low}/\text{Betty} + \text{low}/\text{Helen} + \text{very high}/\text{Ruth} + \text{high}/\text{Pat}, \tag{1.15}$$

where "middle", "low" and "high" are fuzzy sets. So, instead of saying "Helen is beautiful to degree the degree of 0.3", we say that "Helen is beautiful to the degree of *low*".

A simplified form of type-2 sets is known as the interval-valued fuzzy sets. In an interval-valued fuzzy set, the membership degree is defined with an interval in $[0,1]$, refer to Fig. 1.7 for an example. For the recent developments on type-2 fuzzy sets, refer to [134].

1.1.2 Fuzzy Operations

Similar to classical sets, there are some basic operations on fuzzy sets. Consider fuzzy sets A and B on the universe U, whose membership function are $\mu_A(x)$ and $\mu_B(x)$, $\mu_A(x), \mu_B(x) \in [0,1]$, then:

- *Intersection* of A and B, with the notation $A \bigcap B$, is defined by

$$(A \bigcap B)(x) = T(A(x), B(x)), \tag{1.16}$$

  where $T(\cdot)$ is a triangular norm, T-norm for short.
- *Union* of A and B, with the notation $A \bigcup B$, is defined by

$$(A \bigcup B)(x) = T^\star(A(x), B(x)), \tag{1.17}$$

  where $T^\star(\cdot)$ is a T-conorm.
- *Complement* of A, with the notation $\bar{A}$, is defined by

$$\bar{A}(x) = 1 - A(x). \tag{1.18}$$

Many definitions for the T-operators have been given. Before introducing some widely used definitions, we first provide the properties that T-operators must satisfy. Let $T : [0, 1] \times [0, 1] \to [0, 1]$. T is a T-norm if and only if for all $x, y, z \in [0, 1]$:

1. $T(x, y) = T(y, x)$(commutativity),
2. $T(x, y) \leq T(x, z)$, if $y \leq z$(monotonicity),
3. $T(x, T(y, z)) = T(T(x, y), z)$(associativity),
4. $T(x, 1) = x$.

Let $T^\star : [0, 1] \times [0, 1] \to [0, 1]$. $T^\star$ is a T-conorm if and only if for all $x, y, z \in [0, 1]$:

1. $T^\star(x, y) = T^\star(y, x)$(commutativity),
2. $T^\star(x, y) \leq T^\star(x, z)$, if $y \leq z$(monotonicity),
3. $T^\star(x, T^\star(y, z)) = T^\star(T^\star(x, y), z)$(associativity),
4. $T^\star(x, 0) = x$.

Dozens of definitions have been suggested for T-norms and T-conorms. One of the most popular T-norms is the Zadeh operators and are defined as:

$$T_1(x, y) = \min(x, y), \tag{1.19}$$

$$T_1^\star(x, y) = \max(x, y). \tag{1.20}$$

See Fig. 1.8 for a graphic illustration of the operations.

An another pair of most widely used T-operators, which are also called probabilistic operators, are defined as:

$$T_2(x, y) = xy, \tag{1.21}$$

$$T_2^\star(x, y) = x + y - xy. \tag{1.22}$$

Refer to Fig. 1.9 for an illustration of the operators.

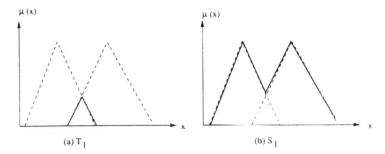

Fig. 1.8. The Zadeh operators.

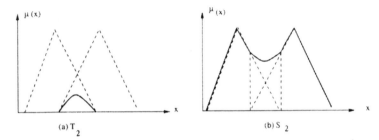

Fig. 1.9. The probabilistic operators.

Different definitions of T-operators have different characteristics. However, they do not exhibit significant discrepancies in fuzzy inference. Interestingly, all definitions of T-operators satisfy the following constraints:

$$T(x,y) \leq \min(x,y), \tag{1.23}$$

$$T^{\star}(x,y) \geq \max(x,y). \tag{1.24}$$

Besides the T-norms, operators that do not satisfy the conditions for T-operators have also been proposed for the fuzzy intersection and union [247]. These operators are called *soft fuzzy operators*. The first pair of soft fuzzy operators are based on the Zadeh operators and defined as follows:

$$\tilde{T}(x,y) = \alpha \cdot x \wedge y + (1-\alpha) \cdot \frac{1}{2}(x+y), \tag{1.25}$$

$$\tilde{T}^{\star}(x,y) = \alpha \cdot x \vee y + (1-\alpha) \cdot \frac{1}{2}(x+y), \tag{1.26}$$

where $0 \leq \alpha \leq 1$, $\wedge$ and $\vee$ denote the minimum and maximum operations, respectively. Consequently, a soft fuzzy intersection operator can also be defined based on the probabilistic fuzzy intersection:

$$\tilde{T}(x,y) = \alpha \cdot xy + (1-\alpha)\frac{1}{2}(x+y). \tag{1.27}$$

Soft fuzzy operators have been proved to have more flexibility than the conventional T-operators because there is a parameter in the soft fuzzy operators, which can be properly tuned for different situations. By adjusting the parameter in the soft T-operators, the performance fuzzy systems can be significantly improved [247, 124]. Optimization of the parameter in the soft fuzzy operators using evolutionary algorithms will be presented in Chapter 6.

1.1.3 Fuzzy Relations

One important facet of the fuzzy set theory is the fuzzy relation. Whereas relation is a general concept among all the things in the world, fuzzy relation is no doubt a better tool to describe whether two different things have any association. Before introducing the mathematical definition for the fuzzy relation, we present a simple example to show what is relation between two elements.

It is well known that weight and height are related to a certain degree. Usually, one is considered healthy if one's weight and height satisfy the following equation:

$$\text{weight}(kg) = \text{height}(cm) - 100. \tag{1.28}$$

To describe this equation in terms of relation, let $H = \{140, 150, 160, 170, 180\}$ be the domain of height, $W = \{40, 50, 60, 70, 80\}$ the domain of weight, and if weight and height satisfy equation (1.28), it is denoted with 1; otherwise it is denoted with 0. Table 1.1 thus gives a description of the relation $R(W, H)$.

Table 1.1. Classical relations between weight and height

R(W,H)	40	50	60	70	80
140	1	0	0	0	0
150	0	1	0	0	0
160	0	0	1	0	0
170	0	0	0	1	0
180	0	0	0	0	1

However, the relation between weight and height can be described more reasonably with fuzzy relations, where a degree between 0 and 1 can be used. In fact, it is hard to satisfy equation (1.28) exactly, and consequently most people will be considered to be unhealthy according to equation (1.28). This situation can be avoided if the fuzzy relation is employed to describe the relationship between the height and weight. Table 1.2 describes the relation between the weight and height with a grade of degree.

Table 1.2. Fuzzy relations between weight and height

R(W,H)	40	50	60	70	80
140	1	0.8	0.2	0.1	0
150	0.8	1	0.8	0.2	0.1
160	0.2	0.8	1	0.8	0.2
170	0.1	0.2	0.8	1	0.8
180	0	0.1	0.2	0.8	1

According to Table 1.2, a person of 170 cm tall he has a perfect weight if he is 70 Kg. However, if he is 80 Kg, the degree of healthy is then 0.8. That is to say, his weight is normal, although not ideal.

Mathematically, a fuzzy relation on $X \times Y$ denoted by $R(x,y)$ is defined as:

$$R(x,y) = \{((x,y), \mu_R(x,y))|(x,y) \in X \times Y, \mu_R(x,y) \in [0,1]\}, \qquad (1.29)$$

where x and y are subsets on the universes X and Y, '$\times$' denotes the Cartesian product, and $\mu_R(x,y)$ is the membership value, which indicates the degree to which x and y are associated.

Since a fuzzy relation is a fuzzy set, operations on fuzzy sets can also be applied to fuzzy relations. For example, assume there are two fuzzy relations R_1 and R_2 on $U = \{a,b\}$, and

$$R_1 = \frac{0.6}{(a,a)} + \frac{0.4}{(a,b)} + \frac{0.3}{(b,a)} + \frac{0.9}{(b,b)}, \qquad (1.30)$$

$$R_2 = \frac{0.8}{(a,a)} + \frac{0.5}{(a,b)} + \frac{0.4}{(b,a)} + \frac{0.2}{(b,b)}. \qquad (1.31)$$

Then,

$$R_1 \bigcap R_2 = \frac{0.6 \wedge 0.8}{(a,a)} + \frac{0.4 \wedge 0.5}{(a,b)} + \frac{0.3 \wedge 0.4}{(b,a)} + \frac{0.9 \wedge 0.2}{(b,b)}$$

$$= \frac{0.6}{(a,a)} + \frac{0.4}{(a,b)} + \frac{0.3}{(b,a)} + \frac{0.2}{(b,b)}. \qquad (1.32)$$

One of the most useful operations on fuzzy relations is the *composition*, which combines fuzzy relations between different variables. Suppose R and S are fuzzy relations in the domain of $U \times V$ and $V \times W$, then the composition of R and S denoted by $R \circ S$ is a fuzzy relation in the domain of $U \times W$, which is defined by:

$$(R \circ S)(u,w) = \vee_{v \in V}(R(u,v) \wedge S(v,w)). \qquad (1.33)$$

Usually, we use the maximum for the fuzzy intersection and the minimum for the fuzzy union. Therefore, this kind of composition is also called the *max-min composition*.

Fuzzy relation composition is very important in fuzzy modeling and control, because it plays an essential role in approximate reasoning. Consider the following question:

If x is A, then y is B;
If x is A', how is y?

To answer this question, we first need to define the fuzzy relation between x and y (also called implication) using fuzzy membership functions:

$$R(x,y) = (A \rightarrow B)(x,y)$$
$$= (A(x) \wedge B(y) \vee (1 - A(x))), \qquad (1.34)$$

where A and B are fuzzy sets on the universe of X and Y, $x \in X$, $y \in Y$, $A(x)$ and $B(y)$ are the membership functions for A and B, respectively. With the help of the fuzzy composition, we can get an answer to the question using approximate reasoning. Suppose y is B' if x is A', then we have:

$$B'(y) = A'(x) \circ R(x,y)$$
$$= \vee_{x \in X}[A'(x) \wedge R(x,y)]$$
$$= \vee_{x \in X}[A'(x) \wedge (A(x) \wedge B(y)) \vee (1 - A(x))]. \qquad (1.35)$$

To illustrate this, we provide the following simple example. Let $X = Y = \{1,2,3,4,5\}$, *Large*, *Small* and *quite Small* are three fuzzy sets on X and Y, which are defined by the following membership functions:

$$Large = (0,0,0.1,0.6,1), \qquad (1.36)$$
$$Small = (1,0.7,0.4,0,0), \qquad (1.37)$$
$$quiteSmall = (1,0.6,0.4,0.2,0). \qquad (1.38)$$

The problem we have is as follows:

If x is *Small*, then y is *Large*;
If x is *quite Small*, how is y?

Using the fuzzy implication defined in equation (1.35), we get:

$$R(x,y) = (Small(x) \wedge Large(y)) \vee (1 - Small(x)) \qquad (1.39)$$

$$= \begin{bmatrix} 0 & 0 & 0.1 & 0.6 & 1 \\ 0.3 & 0.3 & 0.3 & 0.3 & 0.3 \\ 0.6 & 0.6 & 0.6 & 0.6 & 0.6 \\ 1 & 1 & 1 & 1 & 1 \\ 1 & 1 & 1 & 1 & 1 \end{bmatrix}. \qquad (1.40)$$

Given a new premise "x is *quite Small*", we can get y with the help of the fuzzy composition according to equation (1.35):

$$y' = (quiteSmall)(x) \circ R(x,y)$$

$$= (1, 0.6, 0.4, 0.2, 0) \circ \begin{bmatrix} 0 & 0 & 0.1 & 0.6 & 1 \\ 0.3 & 0.3 & 0.3 & 0.3 & 0.3 \\ 0.6 & 0.6 & 0.6 & 0.6 & 0.6 \\ 1 & 1 & 1 & 1 & 1 \\ 1 & 1 & 1 & 1 & 1 \end{bmatrix}$$

$$= (0.4, 0.4, 0.4, 0.6, 1). \tag{1.41}$$

That is, y' is a fuzzy set $\{0.4/1 + 0.4/1 + 0.4/1 + 0.6/4 + 1/5\}$. Compared to the definition of the fuzzy set *Large*, y' can be interpreted as *'quite Large'*. This result is in a good agreement with our intuition.

1.1.4 Measures of Fuzziness

The key to the extension of classical sets to fuzzy sets is the introduction of the membership function. However, a definition for a membership function of a particular fuzzy set is not straight-forward. The determination of fuzzy membership functions is usually a rule of thumb during the earlier time in the development of fuzzy systems, which is a weakness that has often been criticized. Since the end of the 1990's, fuzzy systems have benefited greatly from the techniques developed in the fields of machine learning, artificial neural networks and evolutionary computation. Designing fuzzy systems that can learn is no longer a new concept but a common practice for engineers who design fuzzy systems in a variety of areas. The generation of fuzzy rules and the determination of fuzzy membership functions are dependent more on data collected from experiments than on the human experience only. In this way, the obtained fuzzy rules and fuzzy membership functions are becoming more and more objective.

Nevertheless, no matter how fuzzy membership functions are obtained, it is necessary to develop certain measures to check how 'fuzzy' a fuzzy set is. We will show in the following that a proper fuzziness of a fuzzy set is important for the performance of a fuzzy system. Furthermore, an appropriate fuzziness value is also critical to the interpretability of a fuzzy system. For example, three different membership functions are given in Fig. 1.10 for the fuzzy set *Middle aged*. Your intuition will tell you that the membership functions in Fig. 1.10(a) and Fig. 1.10 (c) are not appropriate, although they are not wrong. This may be ascribed to the improper fuzziness of the membership functions.

The definition of fuzziness, like the definition of fuzzy operators, can also be quite different. However, a measure of fuzziness $F(A) \in [0, 1]$ for any fuzzy set A on the universe U should satisfy the following conditions.

1. $F(A) = 0$, if A is a classical set;
2. $F(A) = 1$, if $A(u) = \frac{1}{2}, \forall u \in U$;

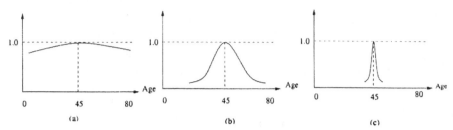

Fig. 1.10. Different fuzzy membership functions for "Middle aged".

3. For A, B on the universe U, if $\sum_{u \in U} |B(u) - \frac{1}{2}| \geq \sum_{u \in U} |A(u) - \frac{1}{2}|$, then $F(B) \leq F(A)$.

These conditions have very clear meanings. The fuzziness is zero if a set is a classical set. A fuzzy set has the maximal degree of fuzziness (fuzziness equals 1) if its membership degree equals 0.5 for all of its elements. Finally, the closer a fuzzy membership function to the $\mu(u) = 0.5$ line, the fuzzier the fuzzy set is.

In order to give the definitions for fuzziness, we first present a measure of distance between two fuzzy sets. The first distance measure is based on the Hamming distance. For two fuzzy sets A and B on the universe $U = \{u_1, u_2, ..., u_n\}$, then

$$D_H(A, B) = \frac{1}{n} \sum_{i=1}^{n} |A(u_i) - B(u_i)| \tag{1.42}$$

is called the distance between fuzzy sets A and B. If universe U is continuous, then the distance is redefined as:

$$D_H(A, B) = \frac{1}{b-a} \int_a^b |A(u) - B(u)| du, \tag{1.43}$$

where a and b are the minimal and maximal elements in the support of fuzzy sets.

If we define the fuzzy distance using the Euclidean distance, the distance definitions can be rewritten as:

$$D_E(A, B) = \sqrt{\frac{1}{n} \sum_{i=1}^{n} |A(u_i) - B(u_i)|^2}, \tag{1.44}$$

$$D_E(A, B) = \sqrt{\frac{1}{b-a} \int_a^b [A(u) - B(u)]^2 du}. \tag{1.45}$$

With the distance measure for fuzzy sets, we can then define a fuzziness measure for a fuzzy set A with the distance to its nearest classical set $A_{0.5}$:

$$F_H(A) = 2D_H = \frac{2}{n} \sum_{i=1}^{n} |A(u_i) - A_{0.5}(u_i)|, \tag{1.46}$$

$$F_E(A) = 2D_E = 2\sqrt{\frac{1}{n} \sum_{i=1}^{n} [A(u_i) - A_{0.5}(u_i)]^2}, \tag{1.47}$$

where $A_{0.5}$ is the nearest classical set of the fuzzy set A defined by:

$$A_{0.5} = \begin{cases} 0 & \text{if } A(u) < 0.5, \\ 1 & \text{if } A(u) \geq 0.5. \end{cases} \tag{1.48}$$

As an example, consider two fuzzy sets A and B on the universe $U = \{u_1, u_2, u_3, u_4, u_5\}$, and let $A = (0.5, 0.4, 0.45, 0.55, 0.6)$, $B = (0, 0.2, 0.4, 0.8, 1)$, then:

$$A_{0.5} = (1, 0, 0, 1, 1), \tag{1.49}$$
$$B_{0.5} = (0, 0, 0, 1, 1), \tag{1.50}$$
$$F_H(A) = 2D_H(A)$$
$$= \frac{2}{5}[|0.5 - 1| + |0.4 - 0| + |0.45 - 0| + |0.55 - 1| + |0.6 - 1|]$$
$$= 0.88, \tag{1.51}$$
$$F_H(B) = 2D_H(B)$$
$$= \frac{2}{5}[|0 - 0| + |0.2 - 0| + |0.4 - 0| + |0.8 - 1| + |1 - 1|]$$
$$= 0.32. \tag{1.52}$$

From the calculated results, it can be seen that the fuzziness of the fuzzy set A is much larger than that of the fuzzy set B. This is consistent with the intuition if we have a look at the fuzzy sets.

1.1.5 Measures of Fuzzy Similarity

It is quite often the case that we need to know how one fuzzy set resembles another. For example, one may use *quite Good* or *not Bad* to express one's opinion on a certain matter. Intuitively, we know that the two linguistic terms are expressing similar meanings. However, actually how similar *quite Good* and *not Bad* are?

Since the properties of a fuzzy set are mainly determined by its membership function, the similarity measure of two fuzzy sets is therefore discussed based on the membership function of the fuzzy sets. A simple similarity measure can be defined using the height of fuzzy sets as follows:

$$S(A, B) = \frac{\text{hgt}(A \cap B)}{\text{hgt}(B)}, \tag{1.53}$$

where A, B are two fuzzy sets, $hgt(\cdot)$ is the height of a fuzzy set as defined in equation (1.11). According to this definition, the fuzzy similarity is a number between $[0, 1]$.

Fuzzy similarity measures can also be defined using distance measures. In the following, we give a similarity measure based on the distance measure:

$$S(A, B) = \frac{1}{1 + D(A, B)}, \tag{1.54}$$

where $D(A, B)$ is the distance between the fuzzy sets A and B. When A equals B, the similarity reaches its maximal value of 1. However, unlike the fuzzy similarity measure based on the height of the fuzzy sets, the fuzzy similarity based on the distance measure is normally larger than zero because the distance between two fuzzy sets can not be infinitive.

Finally, we introduce a fuzzy similarity measure using the size of a fuzzy set:

$$\begin{aligned} S(A, B) &= \frac{M(A \bigcap B)}{M(A \bigcup B)} \\ &= \frac{M(A \bigcap B)}{M(A) + M(B) - M(A \bigcap B)}, \end{aligned} \tag{1.55}$$

where $M(.)$ is the size of a fuzzy set defined as follows:

$$M(A) = \int_{x \in A} A(x) dx. \tag{1.56}$$

The meaning of this definition is obvious. If two fuzzy sets have no overlapping, the similarity measure is 0. If they equal, the fuzzy similarity measure reaches the maximal value of 1.

Although the fuzzy similarity measure based on the size of a fuzzy set is very useful in fuzzy reasoning and fuzzy rule evaluation, the computation of the similarity measure is relatively more complicated, especially when the domain of the fuzzy set is continuous.

1.2 Fuzzy Rule Systems

1.2.1 Linguistic Variables and Linguistic Hedges

A rule system consists of a number of rules with a condition part and an action part:

$$\text{If ¡condition¿, then ¡action¿.}$$

The condition part is also known as the rule premise, or simply the IF part. The action part is also called the consequent part or the THEN part.

A fuzzy rule system is a rule system whose variables or part of its variables are linguistic variables. For example, we have the following rule:

If *error* is *very Small*, then increase *control output slightly*.

In this rule, *error* and *control output* are two linguistic variables, because their values are linguistic terms used in the natural language. The definition of a *linguistic variable* was given by Zadeh as follows:

A *linguistic variable* is a variable whose values are sentences in a natural or artificial language.

According to this definition, if the values of x are *small, middle, large,* or *young, not very young, old,* then x is a linguistic variable. A linguistic variable is characterized by a quintuple $\{x, T(x), G, M, U\}$ in which x is the name of the variable, $T(x)$ is the term set of x, that is, a set of linguistic values of x, which are fuzzy sets on the universe U, G is the syntactic rule for generating the names of values of x, and M is a semantic rule for associating each value with its meaning, that is, the membership function that defines the fuzzy set. Fig. 1.11 describes a linguistic variable *age*. In the figure, *age* is a linguistic variable: *very young, young, middle age, old* are linguistic values. The membership functions for the linguistic terms (fuzzy sets) are called the semantic rules, and the universe U discussed here is $[0, 100]$.

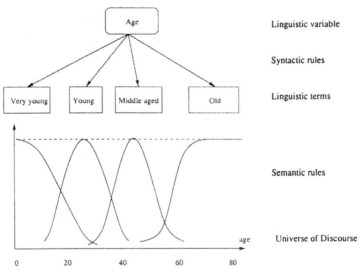

Fig. 1.11. The linguistic variable "age".

From Fig. 1.11, we notice that for a linguistic variable, a number of fuzzy subsets are generated according to the syntactic rule. Generally, we call this collection of fuzzy subsets a *fuzzy partition*. The number of fuzzy sets in a fuzzy partition may be different. For instance, for the linguistic variable *Age*, a fuzzy partition may have three fuzzy sets {*young, middle age, old*}, or five fuzzy subsets {*very young, young, middle age, old, very old*}.

Strictly speaking, a fuzzy partition is a number of fuzzy sets A_1, A_2,...,A_n on the universe of discourse, where $A_i \neq \emptyset, i = 1, 2, ...n$ and $\sum_{i=1}^{n} A_i(u) = 1, \forall u \in U$. However, in most applications, we call $A_i, i = 1, 2, ..., n$ a fuzzy partition even if $\sum_{i=1}^{n} A_i(u) \neq 1$. This will no doubt bring about more flexibility in the design of fuzzy systems. Despite that, attention has also to be paid to the interpretability of the fuzzy systems, about which we will discuss in the following section.

Linguistic values can be modified with linguistic hedges. For instance, *big* can be changed to *very big*, or *quite big* and so on. A linguistic hedge (modifier) may act as an intensifier, such as *very, very very*. If the membership function for x is A_x, where x is a fuzzy set(linguistic term), then the membership function for *very x* can be defined as $(A_x)^2$. In contrast to the linguistic hedges that act as an intensifier, linguistic hedges can also act as a dilator, such as *fairly, more or less* and so on. A dilation of fuzzy set A_x can be defined by $(A_x)^{\frac{1}{2}}$.

The concentration and dilation of linguistic hedges can naturally be extended to powered hedges:

$$m_p(A) = \int_U (\mu_A(u))^p/u, \qquad (1.57)$$

where m_p is the linguistic modifier and p is a parameter. If $0 < p < 1$, m_p is a dilation operation, and if $p > 1$, m_p is a concentration operation. Fig. 1.12 shows two typical cases for the powered hedges.

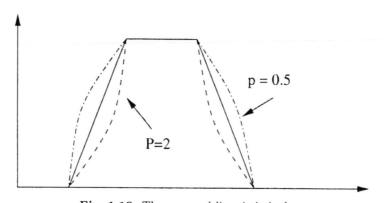

Fig. 1.12. The powered linguistic hedges.

In addition to the powered hedges, *shifted hedges* and *scaled hedges* have also been proposed. The shifted hedges are defined by:

$$m_s(A) = \int_U \mu_A(u - s)/u, \qquad (1.58)$$

where m_s is a linguistic modifier and s represents the magnitude of the shift. Notice that the value of s has different signs within one hedge. For example, if we apply the concentration operation on a fuzzy set, s is positive on the left side of the center of the fuzzy membership function and negative on the right side. For a dilation operation, it is reversed.

To interpret fuzzy rules properly, an appropriate definition of the linguistic hedges is necessary. Unfortunately, most existing definitions for linguistic hedges are still a matter of intuition.

Among the quintuple of the linguistic variables, the definition of a semantic rule, that is, the determination of the membership function for each fuzzy subset is critical for the performance of a fuzzy rule system. Therefore, this has also been one of the challenging topics in the fields of fuzzy rule systems. Generally, determination methods for membership functions can be classified into the following categories:

- Subjective evaluation and heuristics. Since fuzzy sets are usually intended to model the cognitive process of human beings, the membership function of fuzzy sets can be determined based on the experience or intuition of human beings. In practice, certain types of functions, such as the triangular, trapezoidal, Gaussian or bell-shaped functions, are often used. These functions are very easy to use, yet they work effectively in most situations.
- Converted frequencies or probabilities. Membership functions can sometimes be constructed on the basis of frequency histograms or other probability curves. There are a variety of conversion methods, each with its own mathematical and methodological strengths and weaknesses.
- Learning and adaptation. This is the most sophisticated and objective method for determination of membership functions. Based on a set of the training data, parameters of fuzzy membership functions can be learned or adapted using different optimization methods, such as the gradient method, genetic algorithms or reinforcement learning.

1.2.2 Fuzzy Rules for Modeling and Control

Modeling and control is one of the most successful areas in which the fuzzy rule systems have been employed. A simple single-input-single-output system can be described by a fuzzy rule

$$\text{If } x \text{ is } A(x), \text{ then } y \text{ is } B(y),$$

where x is the input, y is the output, and $A(x)$ and $B(y)$ are fuzzy sets defined on the corresponding universe of discourse. By properly defining the fuzzy relation and the fuzzy composition, which is usually called approximate reasoning, a new conclusion $B'(y)$ can be reached given a new premise $A'(x)$:

$$B' = A' \circ R(x, y). \tag{1.59}$$

Besides, a fuzzy rule forms a functional mapping from x to y, which can generally be expressed by $y = f(x)$, where $f(.)$ is determined by the fuzzy relation $R(x, y)$ and the fuzzy composition. Thus, a set of fuzzy rules are able to build up a more complex mapping between two or more variables:

Rule 1: If x_1 is A_{11},....,x_n is A_{1n}, then y is B_1;
Rule 2: If x_1 is A_{21},...,x_n is A_{2n}, then y is B_2;
... Rule N: If x_1 is A_{N1},....,x_n is A_{Nn}, Then y is B_N.

Suppose the fuzzy relation of rule i is R_i, then the whole output of the fuzzy system will be:

$$R = R_1 \bigcup R_2 \bigcup ... \bigcup R_N$$
$$= \bigcup_{i=1}^{N} R_i. \tag{1.60}$$

The definition of the fuzzy relation R_i plays a very important role in the fuzzy inference. Until recently, there has been confusion in the definition of different fuzzy inference mechanisms [143, 51]. This confusion is originated from the difficulty to distinguish between two related but very different fuzzy rules: *fuzzy mapping rules* and *fuzzy implication rules* [250]. A fuzzy mapping rule defines a rough relationship between the input (x) and the output (y), which can be realized by means of a conjunction such as a fuzzy Cartesian product $A \times B$:

$$R_i = (A_{i1} \times A_{i2} \cdots \times A_{in}) \times B_i. \tag{1.61}$$

Thus, the membership function of the fuzzy relation R_i can be defined by

$$\mu_{R_i} = T[T(\mu_{A_{i1}}, \mu_{A_{i2}}, \cdots, \mu_{A_{in}}), \mu_{B_i}]. \tag{1.62}$$

In this sense, fuzzy mapping rules are also called *conjunction-based models* [51]. A collection of fuzzy mapping rules (usually called a *fuzzy model*) can define the functional mapping between the input space and the output space.

The second type fuzzy rules are called *fuzzy implication rules*. Fuzzy implication rules describe a generalized logic implication relationship (denoted by $A \rightarrow B$) between two logic formula involving linguistic variables. Let $t(\cdot)$ the truth value of the implication, there are generally three families of definitions for the inference of fuzzy implication $A \rightarrow B$ [250]:

- The first family of fuzzy implication is obtained by generalizing material implications that define $A \rightarrow B$ as $\neg A \wedge B$ to $t(A \rightarrow B) = t(\neg A \wedge B)$:

$$t(A \rightarrow B) = t(\neg A \wedge B) \tag{1.63}$$
$$= ((1 - A(x)) \oplus B(y). \tag{1.64}$$

- The second family of fuzzy implication is extended from the logic equivalence between $A \rightarrow B$ and $\neg A \vee (A \wedge B)$:

$$t(A \rightarrow B) = t(\neg A \vee (A \wedge B))$$
$$= (1 - A(x)) \oplus (A \otimes B(y)). \tag{1.65}$$

- ·The third family of fuzzy implication is generalized from the "standard sequence" of many-valued logic. In this type of implication, we have $t(A \rightarrow B) = 1$ whenever $t(A) \leq t(B)$, which means that the implication is considered to be true whenever the consequent is as true as or truer than the antecedent [1].

The difference between fuzzy mapping rules and fuzzy implication rules can be attributed to the different knowledge represented by the statement "IF x is A, THEN y is B." If it is interpreted as a fuzzy mapping rule, B is viewed as a lower bound of the possible values of y when x is in A. When the value of x moves from the kernel of A, we can either reduce the lower bound on the level of possibility of the values in B to the degree of membership of x to A, or, we believe that the subset of values with some guaranteed possibility for y becomes smaller and smaller inside B. In contrast, if the rule is interpreted as a fuzzy implication rule, B is viewed as an upper bound of the possible values of y when x is in the kernel of A. Naturally, we modify B into a less restrictive fuzzy set when x moves from its kernel. One way of modification is to attach some uncertainty to B so that the possibility degree of values outside of the support of B is no longer strictly zero. This is to expand the support of B. The other way is to enlarge the kernel of B, i.e., when the value of x is in the α-cut of A ($x \in \{u, A(X) \geq \alpha\}$), the degree of all the values in the α-cut of B increases to 1.

In fuzzy modeling and control, mainly fuzzy mapping rules are used. Fig. 1.13 illustrates a simple example of fuzzy inference for a two-input one-output system. Suppose the fuzzy rule in consideration is

If x is LARGE and y is SMALL, then z is MEDIUM.

See Fig. 1.13 for the fuzzy partitions of x, y and z. Given crisp inputs $x0$ and $y0$, whose membership degrees to "LARGE" and "SMALL" are $\mu_L(x0)$ and $\mu_S(y0)$, respectively, where $\mu_L(x)$ is the membership function of fuzzy set "LARGE" of x, $\mu_S(y)$ is the membership function of fuzzy set "SMALL" of y. Then the truth value of the rule premise is $\mu_S(y0)$, if $\mu_S(y0) < \mu_L(x0)$ and the minimization is used as the T-norm operation. Thus, the contribution of this fuzzy rule to the whole system is the shaded area of the fuzzy membership function "MEDIUM" in Fig. 1.13 (c), which is a fuzzy value.

Generally, the final output of the fuzzy system should be a crisp value. Thus, a defuzzification process is needed to convert the fuzzy output of fuzzy rules to a crisp value. Several schemes for defuzzification have been developed and three most popular ones are mean of maximum, Center of Gravity and center of area.

Suppose for a given input, two fuzzy rules are activated and their contribution is plotted in Fig. 1.14. Thus, different defuzzification methods work as follows.

---

[1] Notice that in fuzzy logic, the truth of a statement is always a degree.

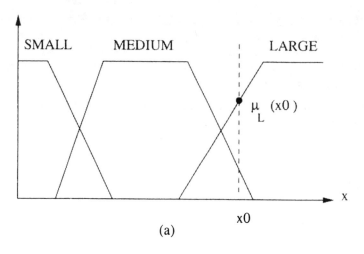

(a)

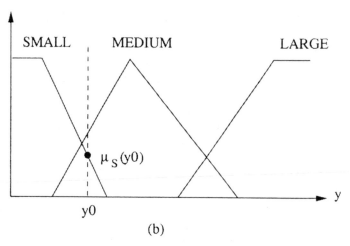

(b)

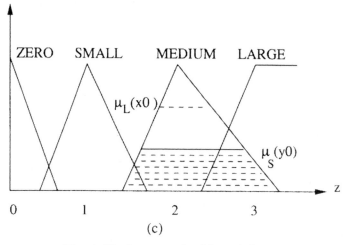

(c)

Fig. 1.13. An example of fuzzy inference.

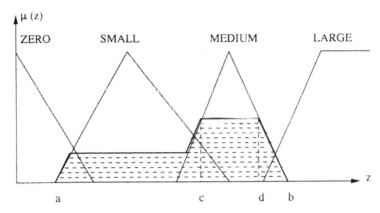

Fig. 1.14. Defuzzification.

- Mean of maximum (MOM). As its name suggests, this method calculates the mean of the values with the maximal membership degree. In this example, the maximal membership degree is reached for $c \leq z \leq d$:

$$z^\star = \frac{\int_{z=c}^{z=d} z\,dz}{\int_{z=c}^{z=d} dz}.$$ (1.66)

- Center of gravity (COG). In this approach, the point where the center of gravity is located is taken as the defuzzified result:

$$z^\star = \frac{\int_{z=a}^{z=b} \mu(z) z\,dz}{\int_{z=a}^{z=b} \mu(z) dz}.$$ (1.67)

- Center of area (COA). This method looks for the point where the shaded area is equally divided:

$$\int_{z=a}^{z=z^\star} \mu(z) dz = \int_{z=z^\star}^{z=b} \mu(z) dz.$$ (1.68)

The MOM method is computationally very simple, however, a lot of information will be lost. Due to this reason, COG and COA are more often used, although the computational complexity is higher. In practice, a simplified defuzzification method is widely used, which will be described in the following sections.

In fuzzy control, fuzzy inference can also be implemented off-line and the results are stored in the form of a look-up table. In this situation, the fuzzy inference process is a little different. At first, the universe of discourse needs to be quantified. Then, the membership function of each fuzzy subset is defined based on the quantified levels. For example, suppose the physical range of variable x is between a and b, where a and b are crisp values. The first thing to do is to determine how many integers are needed to represent the interval $[a, b]$. The finer the quantification, the more exact the inference will be (with the same sets of fuzzy rules). Usually, five to seven levels are used, because a larger number of quantification level will lead to the complication of the look-up table. If seven integers are used to describe the universe of discourse of x, that is, $X = \{-3, -2, -1, 0, +1, +2, +3\}$, then, the continuous interval $[a, b]$ can be mapped to a discrete interval $[-3, +3]$ using the following equation:

$$\bar{x} = (\text{integer})\frac{6 \cdot [x - (a + b)/2]}{b - a}, \tag{1.69}$$

where $(integer)(v)$ means the integer that is closest to the value v. For example, if the physical range of x is $[-6, +6]$ and the current value of x is 5.4, then its corresponding level on the universe is

$$\bar{x} = \frac{(\text{integer})6 \cdot [5.4 - (-6 + 6)/2]}{6 - (-6)}$$
$$= (\text{integer})2.7$$
$$= 3. \tag{1.70}$$

Thus, the memberships of the fuzzy subsets of variable x can be described using Table 1.3.

Table 1.3. Fuzzy membership of variable x

	-3	-2	-1	0	1	2	3
PL	0	0	0	0	0	0.5	1
PS	0	0	0	0	1	0.5	0
ZO	0	0	0.5	1	0.5	0	0
NS	0	0.5	1	0	0	0	0
PL	1	0.5	0	0	0	0	0

A discrete version of the defuzzification process is used in the look-up table fuzzy systems. Suppose the output of the fuzzy system is a fuzzy set defined on a discrete universe $U = [Z_1, Z_n]$:

$$u_f = \mu_1/z_1 + \mu_2/z_2 + \dots + \mu_n/z_n, \tag{1.71}$$

the defuzzification methods discussed above can carried out as follows.

- Mean of maximum. According to this method, the crisp output is obtained by:

$$u_c = z_j, \quad j : \mu_j = \max_{i=1}^{n}\{\mu_i\}. \tag{1.72}$$

If more than one element has the same maximal membership, the average value of these elements is taken as the crisp output.

- Center of gravity. Using this method, the crisp output is the weighted average of all the elements of the fuzzy set:

$$u_c = \frac{\sum_{i=1}^{n} \mu_i \cdot z_i}{\sum_{i=1}^{n} \mu_i}. \tag{1.73}$$

Based on the center of gravity defuzzification method, a more general form of defuzzification can be derived:

$$u_c = \frac{\sum_{i=1}^{n} (\mu_i)^\delta z_i}{\sum_{i=1}^{n} (\mu_i)^\delta}, \tag{1.74}$$

where $\delta \in [0, \infty)$. By properly adjusting δ, it is possible to obtain an optimal defuzzification method for a better performance.

1.2.3 Mamdani Fuzzy Rule Systems

In the field of fuzzy modeling and control, there are mainly two types of fuzzy rules, namely, Mamdani fuzzy rules and Takagi-Sugeno-Kang(TSK) fuzzy rules. In this subsection, we will present a brief introduction to Mamdani fuzzy rule systems.

The Mamdani fuzzy rules can be described in the following form:

R^j: If x_1 is A_1^j and x_2 is A_2^j and ...x_n is A_n^j, then y is B^j, $j = 1, 2, ..., N$,

where R^j is the label of j-th rule, $x_i, i = 1, 2, ..., n$ are the inputs of the fuzzy system, $A_i^j(i = 1, 2, ..., n; j = 1, 2, ..., N)$, and B^j are fuzzy subsets for input and output, respectively. Without the loss of generality, we suppose all fuzzy subsets are normal, n is the number of input variables, and N is the number of fuzzy rules. According to the Mamdani fuzzy implication method, the fuzzy relation of the j-th rule can be expressed by:

$$R^j = (A_1^j \times A_2^j \times ... \times A_n^j) \times B^j. \tag{1.75}$$

Rewrite the fuzzy relation with membership function:

$$\mu_{R^j}(x_1, x_2, ..., x_n, y) = \mu_{A_1^j} \wedge \mu_{A_1^j} \wedge \cdots \wedge \mu_{A_n^j} \wedge \mu_{B^j}, \tag{1.76}$$

where $\wedge$ denotes the minimum operator. The overall fuzzy relation of the fuzzy system(R) can be aggregated as follows:

$$\mu_R(x_1, x_2, ..., x_n, y) = \bigvee_{j=1}^{N} [\mu_{R^j}(x_1, x_2, ..., x_n, y)], \tag{1.77}$$

where $\vee$ is the maximum operator.

For the sake of simplicity, the following defuzzification is used:

$$y_c = \frac{\sum_{j=1}^{N} m^j \cdot y^j}{\sum_{j=1}^{N} m^j}, \tag{1.78}$$

where y^j is the kernel of fuzzy set $B^j(y)$, and m_j is the truth value of the premise of each fuzzy rule:

$$m^j = \wedge_{i=1}^{n} A_i^j(x_i). \tag{1.79}$$

If the kernel of the fuzzy set consists of more than one element, then y^j is the mean value of the kernel.

1.2.4 Takagi-Sugeno-Kang Fuzzy Rule Systems

Takagi-Sugeno-Kang(TSK) fuzzy rule systems are receiving more and more attention in the recent years. The main difference between TSK models and Mamdani models lies in the fact that the consequent part of TSK fuzzy rules is a real-valued function of the input variables instead of a fuzzy set. Due to this feature, TSK fuzzy rule systems have the following merits over the Mamdani fuzzy rule systems:

- TSK rule systems are more suitable for various kind of learning algorithms.
- TSK rule systems have stronger representative power and therefore are capable of dealing with complex systems.

TSK fuzzy rules can be expressed in the following general form:

$$R^j: \text{If } x_1 \text{ is } A_1^j, \text{ and..., and } x_n \text{ is } A_n^j, \text{ then}$$
$$y = f(x_1, x_2, ..., x_n),$$

where $f(.)$ is a real-valued function. Typically, TSK fuzzy rule systems have the following different types:

- *Zero-order TSK model.* If the function in the consequent part of the fuzzy rule is a constant, the fuzzy system is called *Zero-order TSK model:*

$$R^j: \text{If } x_1 \text{ is } A_1^j \text{ and ... and } x_n \text{ is } A_n^j, \text{ then } y = c^j,$$

where c^j is a crisp constant value.
- *First-order TSK model.* If function $f(.)$ is a linear function of the input variables, the resulting TSK fuzzy model is called *First-order TSK model,* which is the most commonly used model in fuzzy modeling and control, especially in adaptive fuzzy systems:

$$R^j: \text{If } x_1 \text{ is } A_1^j \text{ and ... and } x_n \text{ is } A_n^j,$$
$$\text{then } y = c_0^j + c_1^j x_1 + ... + c_n^j x_n.$$

The overall output of the TSK model can be computed as

$$y = \frac{\sum_{j=1}^{N} w^j \cdot f(x_1, ..., x_n)}{\sum_{j=1}^{N} w^j}, \tag{1.80}$$

where w^j can be seen as the firing strength of rule R^j

$$w^j = T_{i=1}^n A_i^j(x_i), \qquad (1.81)$$

where T stands for a T-norm.

TSK models are very useful for the modeling and control of nonlinear systems. With these models, a non-linear system can be converted into a set of linear models, which makes it easier to carry out concrete theoretical analyses and to design controllers with the help of the sophisticated theories developed in the field of linear systems. For example, the following fuzzy rule system can describe a nonlinear system to an arbitrary degree:

If x_1 is L_1, $\cdots$, x_n is L_n, then $Y = AX + BU$,

where $X = [x_1, x_2, ..., x_n]^T$ is the system state vector, Y and U are the system output and the control vector, A and B are the transfer matrix and control matrix. $L_i(i = 1, 2, ..., n)$ are fuzzy subsets for the system state x_i. It has been shown that a fuzzy system consisting of the above fuzzy linear sub-systems are capable of describing a highly nonlinear system.

1.2.5 Fuzzy Systems are Universal Approximators

When a fuzzy system is designed for modeling and control, the most important thing is that the designed fuzzy system is theoretically able to realize the desired functional mapping. Therefore, the approximation capability of fuzzy systems is of great concern. It has been shown that both the Mamdani-type and the TSK-type fuzzy rule systems are universal approximators. Furthermore, it has also been shown that various types of commonly used membership functions satisfy the conditions for the fuzzy systems to be universal approximators.

By universal approximators, we mean that a fuzzy system can approximate any continuous functions on a compact set to an arbitrary degree of accuracy. Using more formal mathematical terms, the universal approximation theory can be stated as follows.

Universal approximation property. Let $g(x)$ be a given continuous function and X is compact(closed and bounded). Then for an arbitrary real number $\epsilon > 0$, there exists a fuzzy system $f(x)$, such that

$$\sup_{x \in X} |g(x) - f(x)| < \epsilon, \qquad (1.82)$$

where $f(x)$ is a fuzzy system in the following form:

$$f(x) = \frac{\sum_{i=1}^N A_i y_i}{\sum_{i=1}^N A_i}. \qquad (1.83)$$

It is obvious that the fuzzy system in Equation (1.83) is a widely used expression for the following Mamdani-type single-input single-output fuzzy rule system:

$$R_i\text{:If } x \text{ is } A_i, \text{ then } y \text{ is } B_i,$$

where $A_i, i = 1, 2, ..., N$ are fuzzy membership functions for x, B_i are fuzzy membership functions for y. Note that y_i in Equation (1.83) is the point where $B_i(y)$ reaches the maximum. Usually, all the fuzzy subsets are supposed to be normal and therefore, y_i is the kernel of the fuzzy set.

The proof of the universal approximation based mainly on the Stone-Weierstrass theorem popular in functional analysis and the conclusion can be extended to multiple-input single output systems [49].

The membership functions $A_i(x), i = 1, 2, ..., N$ can be extended to fuzzy basis functions (FBF). A fuzzy basis function is supposed to be complete, consistent and normal. Note that in this context, fuzzy sets $A_i, i = 1, 2, ..., N$ are said to be consistent if $A_i(x_0) = 1$ for some $x_0 \in X$, then for all $j \neq i$, $A_j(x_0) = 0$. Obviously, the fuzzy subsets of a strict fuzzy partition satisfy this condition.

Similarly, it has also been proved that the first-order TSK fuzzy systems are universal approximators. Consider the following first-order TSK model:

$$R_i\text{:If } x \text{ is } A_i, \text{ then } y = a_i + b_i x.$$

The TSK rules can be rewritten as

$$R_i\text{: If } x \text{ is } A_i, \text{ then } y = k_i(a + bx),$$

where

$$a_i = k_i a, \tag{1.84}$$

$$b_i = k_i b. \tag{1.85}$$

The output of the TSK model can be given by:

$$f(x) = \frac{\sum_{i=1}^{N}(A_i(x))^\alpha k_i(a + bx)}{\sum_{i=1}^{N}(A_i(x))^\alpha}, \tag{1.86}$$

where $\alpha \in [0, +\infty)$. Therefore, this is a generalized defuzzification form. Specially when $\alpha = 1$, it is the most popular centroid defuzzifier and when $\alpha \to \infty$, it is the mean of maximum method.

The proof can be carried out in two phases [252]. Firstly, the fuzzy system in Equation (1.86) is shown to be able to approximate, to an arbitrary accuracy, any polynomials. Then, it is proved that the polynomials can uniformly approximate a continuous function $g(x)$ with an arbitrary accuracy using the Weierstrass theorem.

Sufficient conditions for the universal approximation have also been given [252]. Given a desired accuracy ϵ, the minimal number N of fuzzy rules needed to approximate a given function with the prescribed accuracy:

$$N = n^\star$$

$$> \frac{|\beta_1| + \sum_{i=1}^{h} |\beta_i|(2^i - 1)}{\epsilon_2}, \tag{1.87}$$

where $\beta_i, i = 1, 2, ..., h$ are the coefficients of the polynomial that can approximate the given function $g(x)$, h is the order of the polynomial, ϵ_2 is the accuracy that the fuzzy system can approximate the polynomial. For example, the given function is $g(x) = e^x$ defined on $[-1, 1]$. If we intend to use a first order TSK model to approximate $g(x)$ with an accuracy of 0.1, how many fuzzy rules are needed?

At first, use the following polynomials to approximate the given function

$$e^x \approx 1 + x + \frac{x^2}{2} + \frac{x^3}{6} + \frac{x^4}{24} + \frac{x^5}{120} \tag{1.88}$$

with an accuracy of about 0.0038. Then, in order that the fuzzy system can approximate e^x with an accuracy of 0.1, it must approximate the polynomials with an accuracy of $0.1 - 0.0038 = 0.0962$. Thus, we have $h = 5$, $\beta_1 = 1, \beta_2 = \frac{1}{2}, \beta_3 = \frac{1}{6}$, $\beta_4 = \frac{1}{24}$, and $\beta_5 = \frac{1}{120}$, for $\epsilon_2 = 0.0962, n^* = 57.7$. Consequently, the minimal needed rule number $N = 2n^* = 117$. If the prescribed accuracy is 0.01, then the minimal number of rules will be 1793.

It should be pointed out that the rule number calculated according to Equation (1.87) is very likely conservative. In fact, a lower bound that is much smaller can be established. Besides, the actual number of required rules depends also on the membership functions and the defuzzification method.

1.3 Interpretability of Fuzzy Rule System

1.3.1 Introduction

Fuzzy systems are believed to be suitable for knowledge representation and knowledge processing. One of the most important reasons to use fuzzy systems for modeling and control is that fuzzy rules are comprehensible for human beings. This enables the fuzzy model to take advantage of the *a priori* knowledge, on the other hand, the resulting model can be properly interpreted and thus evaluated.

Currently, there is no formal definitions for the interpretability of fuzzy systems. In this section, several aspects that are believed to be essential for the interpretability of fuzzy systems will be discussed [125, 229, 35, 114]. They include the properties of fuzzy membership functions, of the fuzzy partitions of the linguistic variable, of the consistency of the rule base, and of the structure of the rule base. For the fuzzy partition of the linguistic variables, both the completeness and distinguishability are considered. Furthermore, the number of fuzzy subsets in a partition should be limited (empirically not larger than ten). Completeness and compactness will be discussed with regard to the rule structure.

1.3.2 The Properties of Membership Functions

The properties of the fuzzy membership functions are believed to be essential for the interpretability of fuzzy systems due to the fact that a fuzzy set needs to satisfy certain conditions so that it can be associated with a semantic meaning. Among others, unimodality and normality are two important aspects [35]. Fortunately, the most widely used membership functions, e.g., the Gaussian membership functions satisfy these properties. On the other hand, if the universe of discourse is $U = [a, b]$, then the leftmost membership function $A_1(x)$ should assume its maximum membership degree at $x = a$ and its rightmost membership function $A_M(x)$ its maximum membership degree at $x = b$, where M is the number of fuzzy subset in the fuzzy partition. In other words, $A_1(a) = A_M(b) = 1$, taking the normality condition into account.

One additional aspect that should be taken into account is the fuzziness of the fuzzy sets. As previously discussed, the fuzziness of a fuzzy set may vary a lot given different membership functions. If a fuzzy set is normal, the fuzzy set will reduce to a crisp interval when the size of the fuzzy set is too large and a singleton when the size decreases to zero, refer to Fig. 1.10 (a) and (c). In both cases, the interpretability of the fuzzy membership function is not good, because they give rise to difficulties in assigning a linguistic term to them.

1.3.3 Completeness of Fuzzy Partitions

The completeness of a fuzzy rule system is determined by the completeness of the fuzzy partition of all input variables and the completeness of the rule structure. We first discuss the completeness of the fuzzy partitions. Suppose an input variable x is partitioned into M fuzzy subsets $A_i(x), i = 1, 2, ..., M$, then the partition is said to be complete if the following condition holds:

$$\forall_{x \in U} \exists_{1 \leq i \leq M} A_i(x) > 0. \tag{1.89}$$

Furthermore, a measure of completeness for a fuzzy partition can be defined by

$$CM(x) = \sum_{i=1}^{M} A_i(x). \tag{1.90}$$

If $CM(x) = 1$ for all x, the fuzzy subsets construct a fuzzy partition in a strict sense and therefore is said to be strictly complete. If $CM(x) = 0$, then the fuzzy partition is incomplete.

Similarly, a measure of completeness for the rule base can also be defined. Assume the rule base has N fuzzy rules and each rule has n antecedents, which are connected by T-norm T, then the completeness of the fuzzy rule system is given by:

$$CM(R) = \sum_{j=1}^{N} T_{i=1}^{n} A_{i,j}(x_i). \tag{1.91}$$

It is obvious that the completeness measure varies on different points in the input space. However, if there exists a point where the completeness measure equals zero, the fuzzy system is said to be incomplete.

Incompleteness of fuzzy control systems may result in unpredictable actions, which is very undesirable in system control. Several methods can be used to address the incompleteness of fuzzy partitions:

- Repairing. Repairing is the most naive method for guaranteeing the completeness of the fuzzy partitions. It is noticed that the fuzzy partition in Fig. 1.15 is incomplete. One approach to repairing the partition is to modify parameters c_1 and a_2 until $c_1 > a_2$.

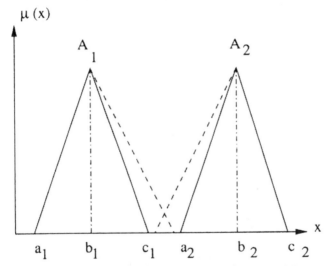

Fig. 1.15. Repairing of an incomplete fuzzy partition.

- Interpolation. Suppose a fuzzy rule base has the following two rules:

$$R_1: \text{If } x \text{ is } A_1, \text{ then } y \text{ is } B_1;$$
$$R_2: \text{If } x \text{ is } A_2, \text{ then } y \text{ is } B_2.$$

The fuzzy membership functions are illustrated in Fig. 1.16(a) and (b). To make the rule base complete, a third fuzzy rule

$$R_3: \text{If } x \text{ is } A', \text{ then } y \text{ is } B'$$

can be generated. The key issue is how to determine parameters of the fuzzy membership functions A' and B'. A general condition that the fuzzy rules are consistent. Therefore, a straightforward method is to use the average of the parameters of A_1 and A_2 to construct A'. Alternatively, the parameters of A' can be determined as follows:

$$a' = b_1, \tag{1.92}$$

$$b' = \frac{b_1 + b_2}{2}, \tag{1.93}$$

$$c' = b_2. \tag{1.94}$$

Similar methods can be developed if the membership functions are trapezoidal.

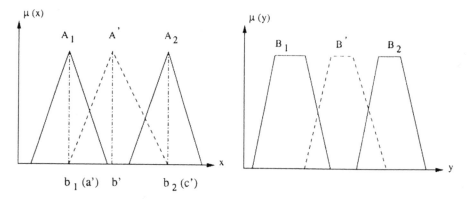

Fig. 1.16. Construction of complete fuzzy partitions by interpolation.

- Constraints. One popular method for avoiding incompleteness in the adaption of fuzzy systems is to introduce constraints on the fuzzy subsets in a fuzzy partition. Suppose a fuzzy partition is defined by a number of triangular fuzzy sets:

$$A_i(x) = \begin{cases} 0 & , \text{ if } x < a_i \\ \frac{x - a_i}{b_i - a_i} & , \text{ if } a_i \le x \le b_i \\ \frac{x - c_i}{b_i - c_i} & , \text{ if } b_i < x \le c_i \\ 0 & , \text{ if } x > c_i \end{cases} \tag{1.95}$$

where $a_i < b_i < c_i$. If the following constraints are satisfied, the fuzzy partition is always complete:

$$a_{i+1} = b_i \tag{1.96}$$

$$b_{i+1} = c_i, \tag{1.97}$$

where A_i and A_{i+1} are two neighboring fuzzy subsets in the fuzzy partition, as illustrated in Fig. 1.17.

Other constraint methods can also be used to guarantee the completeness of the fuzzy partitions. For example, to limit the range of the parameters, or to maintain a predefined order of the parameters, e.g., $c_i > a_{i+1}$.

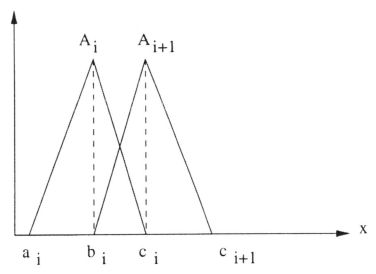

Fig. 1.17. Two neighboring fuzzy sets in a fuzzy partition.

Besides the hard constraint methods described above, soft constraints can also be applied during optimization of membership functions. One method based on the definition of the completeness will be described in Chapter 7.

1.3.4 Distinguishability of Fuzzy Partitions

Distinguishability is one of the main requirements for interpretability of fuzzy systems. In designing an interpretable fuzzy system, one usually assigns a linguistic term to a fuzzy subset. However, there is a prerequisite for this, that is, the fuzzy subsets in a partition should be distinguishable. In Fig.1.18 (a), there are four well distinguishable fuzzy subsets, which can be labeled {ZERO, SMALL, MEDIUM, LARGE}. However, the eight fuzzy subsets in Fig.1.18(b) cannot easily be distinguished. In this case, it is hard to assign an understandable linguistic term to each fuzzy subset and thus the fuzzy rules described by such fuzzy sets are hard to understand.

To improve the distinguishability of fuzzy partitions, one can merge some of the similar fuzzy subsets, especially in adaptive fuzzy systems [205]. However, hard merging of fuzzy subsets may seriously degrade the performance of the fuzzy system. Therefore, an adaptive merging of subsets is suggested in [114].

The distinguishability of a fuzzy partition depends also on the number of fuzzy subsets within a fuzzy partition, refer to Fig.1.18(b). The smaller the number of fuzzy subsets in a fuzzy partition, the more distinguishable the fuzzy partition is, provided that it is complete. In fact, if a fuzzy rule system is designed based on expert knowledge or common sense, the number of fuzzy subset one uses is usually smaller than ten [229].

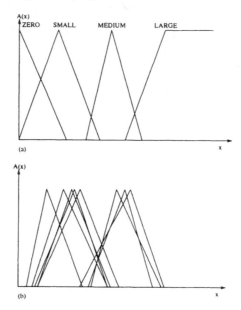

Fig. 1.18. Fuzzy partitions: (a) Distinguishable; (b) Indistinguishable.

1.3.5 Consistency of Fuzzy Rules

Consistency of the rules in a fuzzy rule base concerns with the consistency of the validness of the knowledge represented by the rule base.

Roughly speaking, fuzzy rules are conflicting to each other in the following situations:

1. The condition parts are the same, but the consequent parts are completely different. For example,

   > R1: If x_1 is A_1 and x_2 is A_2, then y is positive Large;
   > R2: If x_1 is A_1 and x_2 is A_2, then y is negative Large.

2. Although the condition parts are seemingly different, but they are the same in physical meanings. However, the consequent parts of the rules are completely different.

   > R1: If x_1 is A_1 and x_2 is A_2, then y is positive Large;
   > R2: If x_1 is A_1 and x_3 is A_3, then y is negative Large.

   Although "x_2 is A_2" and "x_3 is A_3" appears to be different conditions, they may imply the same situation in some cases. For example, for a chemical reactor, a statement "If temperature is high" may imply "conversion rate is high".

3. The conditions in the rule premise are contradictory, e.g. "If x_1 is A_1 and x_2 is A_2,then y is B. However, " x_1 is A_1" contradicts to "x_2 is A_2".

4. The actions in the rule consequent part are contradictory. For example, in the rule "If x is A then y is B and z is C. However, "y is B and z is C" cannot happen simultaneously.

Let us now discuss case (1) in more detail. It is obvious that the two fuzzy rules are inconsistent because they have the same premise part, but their consequent part is completely different.

R_1: If x is positive Small, then y is positive Large;
R_2: If x is positive Small, then y is negative Large.

However, two fuzzy rules can be inconsistent even if their premises are not the same. For example, consider the following two rules:

R_1: If x is positive Small, Then y is positive Large;
R_2: If x is positive Medium, then y is negative Large.

In these two rules, neither the premise part nor the consequent part is the same. Nevertheless, these two rules sound quite inconsistent because their premise parts are quite similar, but their consequent parts are completely different. However, if we rewrite the rules as:

R_1: If x is positive Small, Then y is positive Large;
R_2: If x is positive Medium, Then y is positive Medium.

If the premise parts of two fuzzy rules are completely different, then it is hard to say that they are inconsistent, no matter what consequents of the two rules are. For example:

R_1: If x is positive Large, Then y is positive Small;
R_2: If x is negative Large, Then y is positive Small.

In these two rules, their premise parts are completely different, although their consequent is the same. However, one cannot say that the two rules are inconsistent.

From the previous three simple examples, we make the following observations concerning the consistency of two fuzzy rules.

- If two fuzzy rules have the same premise, but different consequent parts, they are inconsistent.
- If the premise parts of two rules are significantly different, they are always considered to be consistent, no matter what consequents they have.
- If two rules have similar premises and similar consequents, they are also consistent.

According to these observations, a consistency measure for fuzzy rules has been developed using a fuzzy similarity measure [126]. Consider the following two fuzzy rules:

R_i: If x_1 is A_{i1}, and ... and x_n is A_{in}, then y is B_i;
R_k: If x_1 is A_{k1} and ... and x_n is A_{kn}, then y is B_k.

We first define *the similarity of rule premise* (SRP) by:

$$SRP(R_i, R_k) = \min_{j=1}^n S(A_{ij}, A_{kj}),\qquad (1.98)$$

where $S(A_{ij}, A_{kj})$ is the fuzzy similarity between the fuzzy sets A_{ij} and A_{kj}. We can define the fuzzy similarity measure as [159]:

$$S(A_{ij}, A_{kj}) = \frac{M(A_{ij} \cap A_{kj})}{M(A_{ij}) + M(A_{kj}) - M(A_{ij}, \cap A_{kj})}\qquad (1.99)$$

where $M(A)$ is the size of the fuzzy set A.

Following the similarity of rule premise, we also define a *similarity of rule consequent* (SRC) as follows:

$$SRC(R_i, R_k) = S(B_i, B_k),\qquad (1.100)$$

where $S(B_i, B_k)$ is the similarity between the fuzzy sets B_i and B_k. Thus, the consistency of the two fuzzy rules can be defined by:

$$CN(R_i, R_k) = \exp\left\{ -\frac{(\frac{SRP(R_i,R_k)}{SRC(R_i,R_k)} - 1.0)^2}{(\frac{1}{SRP(R_i,R_k)})^2} \right\}.\qquad (1.101)$$

This definition for the consistency of fuzzy rules has two fundamental properties. First, the degree of the consistency tends to be high when the SRP and SRC of the two rules are in proportion, provided that the SRP of the two rules is high. In the special case, if the rules have the same premise and the same consequent, the consistency reaches the highest value of 1. When the premises are the same but the consequent parts are different, the consistency degree varies from 0 to 1.0. Furthermore, if the premise parts of the two rules have low similarity, in other words, the SRP is very small, then the degree of consistency is always high, no matter how the SRC changes. This agrees with our previous observations. One additional remark on the consistency definition is that it is mainly suitable for the Mamdani-type fuzzy rules. For the Takagi-Sugeno-Kang (TSK) fuzzy rules, the consistency is harder to evaluate, when the rule consequent is a function of the input variables. However, the interpretability of the TSK fuzzy rules may be investigated in terms of the physical meaning of each local model that the rule consequent carries [251].

Consistency evaluation of fuzzy rules is sometimes difficult. This happens when a fuzzy system has more than one output variable and the relationship between these variables is unclear. Consider the following two fuzzy rules:

R_1: If x is Small, then y is Large;
R_2: If x is Small, then z is Small.

If we do not consider the relationship between y and z, then these two fuzzy rules are not contradictory. However, if y and z has some inherent relationship, say, if y is *Large*, z cannot be *Small*, then, these two rules are seriously inconsistent. In this case, more careful analysis is required.

1.3.6 Completeness and Compactness of Rule Structure

As a matter of fact, the completeness of fuzzy partitions does not guarantee the completeness of the fuzzy system if the rule structure is not complete. What is the completeness of fuzzy rule structure? A straightforward answer is that all subsets within the fuzzy partition of each input should be used by the fuzzy rule system at least by once. The most widely used full grid rule structure is complete, as shown in Fig. 1.19(a).

However, a complete rule structure is not necessarily a full grid rule structure. As shown in Fig. 1.19(b), the rule structure is complete, although there is no rule in some of the elements. On the contrary, the rule structure in 1.19(c) is incomplete. In fact, a compact but complete fuzzy rule structure is not only feasible, but also very essential, eespecially for large systems.

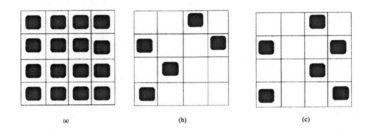

Fig. 1.19. Rule structures. A shaded area denotes a fuzzy rule. (a) A standard grid rule structure, (b) a non-standard but complete rule structure, and (c) an incomplete rule structure.

The compactness of the rule structure concerns mainly with the number of fuzzy variables that appear in the rule premises. If a fuzzy system has 10 inputs and all the inputs appear in the premises of the fuzzy rules, it is very difficult to understand the rules. One effective method for reducing the number of variables appearing in the rule premise is to optimize the rule structure with evolutionary methods [113, 126]. Besides, the compactness of the rule base, in other words, the number of fuzzy rules in the rule base also plays an important role in the interpretability of a fuzzy system. The smaller the number of fuzzy rules in the rule base, the easier for the human user to understand the rule system.

1.4 Knowledge Processing with Fuzzy Logic

1.4.1 Knowledge Representation and Acquisition with IF-THEN Rules

Usually, systems that can process knowledge are called knowledge-based systems. One of the most popular and successful knowledge-based systems is the expert system. Roughly, an expert system is composed of an inference mechanism, a knowledge base, usually in the forms of rules representing specific domain knowledge, and a data base. There are several types of knowledge representation, such as the logical knowledge representation, the procedural knowledge representation, the network knowledge representation and the structured knowledge representation. In the first case, knowledge is represented by an expression in formal logic, which can be extended to fuzzy logic. In the second case, knowledge is described by a set of instructions, for example, if-then rules. A rule can be interpreted as a procedure that achieves a goal for a given premise. Fuzzy if-then rule systems are a large class of rule-based systems. Both the network and the structured knowledge representation schemes represent knowledge using graphs [46].

Fuzzy if-then rule systems are most widely used in fuzzy knowledge representation and processing. A typical fuzzy rule system consists of a set of rules as follows:

$$\text{If } x = A_1, \text{ then } y = B_1;$$
$$\text{If } x = A_2, \text{ then } y = B_2;$$
$$...;$$
$$\text{If } x = A_n, \text{ then } y = B_n.$$

In the fuzzy rules, $A_i, B_i, i = 1, 2, ..., n$ are fuzzy linguistic terms. For the definition and determination of the linguistic variables, please refer to the corresponding section of this chapter.

One of the most popular examples of fuzzy rule-based controllers is to determine the change of control output according to the error and the change of error of the system. If we define:

$$e(t) = y_d(t) - y(t), \tag{1.102}$$
$$\Delta e(t) = e(t) - e(t-1), \tag{1.103}$$
$$\Delta u(t) = u(t) - u(t-1), \tag{1.104}$$

where y and y_d are real and the desired system outputs, $e(t)$ is the error, $\Delta e(t)$ is the change of error, $u(t)$ and $\Delta u(t)$ are the control and the change of control. For such systems, it is straightforward for a control expert to gain the following knowledge:

If the system output is a little above the desired output and is gradually approaching the desired system output, then keep the control unchanged.

Let us interpret this piece of knowledge further. At first, the system output is a little above the desired output means that the system error $e(t)$ is negative small. Since the system output is approaching the desired point gradually, it indicates that the system error is decreasing, i.e., the change of error is positive small as previously defined. In this case, the current control is fairly good and needs not to be changed. Thus, we can represent the knowledge with the following fuzzy rule.

If $e(t)$ is NS, $\Delta e(t)$ is PS, then $\Delta u(t)$ is ZO,

where NS, ZO, PS stand for *negative Small*, *Zero*, and *positive Small*. In practice, it is necessary to convert the variables of the fuzzy controller to proper linguistic variables based on the related domain knowledge.

Several methods have been used for acquiring knowledge for fuzzy rule base systems. In general, knowledge can be achieved from human experts or from experimental data. Mainly, there are three different approaches:

- *Indirect Knowledge Acquisition.* When one designs a knowledge-based system, one can talk to an expert or an experienced operator to gather necessary knowledge.
- *Direct Knowledge Acquisition.* In this case, the designer himself is an expert. The designer not only has rich knowledge in the related field, but is also able to formulate his knowledge in a proper fashion so that it reflects the system correctly.
- *Automatic Knowledge Acquisition.* Most automatic knowledge acquisition methods are developed in the field of machine learning and artificial intelligence.

Those methods use a particular representation, such as semantic networks, conceptual graphs, production rules, fuzzy rules and artificial neural networks. Recently, data mining, that is, to abstract some understandable knowledge from database, has been attracting more and more attention. A combination of neural networks, fuzzy systems and evolutionary computation provides a promising way for automatic knowledge acquisition.

There are two important issues in automatic knowledge acquisition. First, the interpretability of the acquired knowledge should be guaranteed. In other words, the knowledge acquired automatically is not necessarily easily understandable. Second, the knowledge acquired on the basis of experimental data needs to be validated. It should be pointed out that the data may contain redundant, inconsistent and incomplete or even wrong information.

An Example of Heuristic Fuzzy Rule Design. Although data-based automatic fuzzy rule system design is becoming more and more feasible, heuristic fuzzy rule system design is still very important. On the one hand, some kind of knowledge is only available in the form of common sense, intuition and experience. In some cases, the generation of experimental data may be very expensive. On the other hand, even if sufficient data are available for automatic fuzzy rule generation, it is sometimes necessary to build up an initial fuzzy model and then to optimize the model using the experimental data. In the following, we give an example of heuristic fuzzy rule design.

Suppose we are driving a car. It is intuitive that whether to speed up or slow down the car is dependent on the following factors:

- *Safety distance.* When following a car, each driver has a concept of safety distance. Usually, this distance depends on the speed of the cars, the road

conditions and the driving behavior. As a rule of thumb, it is often suggested to be one car length back for each ten miles per hour of the speed.
- *Relative speed.* Relative speed means the speed difference between your car and the car you are following.
- *Traffic regulations and signs.* These include the speed limit and other traffic signs such as stop, yield, traffic light and so on.

The goal of the fuzzy system is to control the car properly so that it is able to follow the car ahead of you properly. For the sake of simplicity, we only consider the two most essential factors, namely, the safety distance and the relative speed in designing the fuzzy rule system. The design procedure can generally be divided into the following steps:

1. Determination of the input and the output. In this case, we select the safety distance (d_s) and the relative speed $(v_r = v_1 - v_2)$ as the inputs of the fuzzy system and the *Acceleration(a)* as the output of the fuzzy system, where v_1 is the speed of the car to be followed, v_2 is the speed of the car to be controlled. For convenience, we use a normalized safety distance (d_n) instead of the safety distance (d_s) itself.
2. Determination the linguistic terms and their corresponding fuzzy membership functions. As we introduced in the previous section, it is necessary to determine the universe of discourse, the number of fuzzy subsets in the fuzzy partitions and the fuzzy membership function for each fuzzy subset. Without the loss of generality, we define the following linguistic variables:

$$d_n = \{\text{Very Small, Small, OK, Large, Very Large }\},$$
$$v_r = \{\text{Negative Large, Negative Small, Zero,}$$
$$\text{Positive Small, Positive Large }\},$$
$$a = \{\text{Negative Large, Negative Medium, Negative Small, Zero,}$$
$$\text{Positive Small, Positive Large }\}. \qquad (1.105)$$

Some additional remarks may be necessary on the linguistic terms of the variables. For the normalized safety distance, *OK* means the distance is about the recommended safety distance, and consequently, *Very Small* indicates that the distance is much closer than the recommended safety distance. On the contrary, *Very Large* says that the distance is much longer than the safety distance. The meaning of the linguistic terms of the relative speed is obvious. As previously defined, a positive relative speed represents the situation in which the car ahead of the driver is faster than the car to be controlled, which implies that the distance between the two cars is increasing. If the relative speed is negative, the distance between the two cars is reducing. It is straightforward that a negative acceleration slows down the car whereas a positive one speeds up the car. Usually, triangular, Gaussian or trapezoidal membership functions can be adopted. Without the loss of generality, triangular membership functions are used here. Assume the universes of discourse of d_n and v_r are $[-1, 1]$ and $[-30, 30]$, respectively, the membership functions for d_n and v_r are given in Fig. 1.20 and Fig. 1.21, respectively. For the fuzzy system output a, its membership functions are shown in Fig. 1.22.

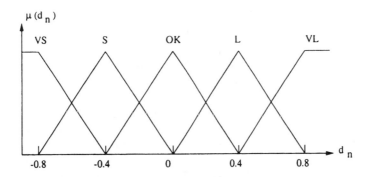

Fig. 1.20. Membership functions of d_n.

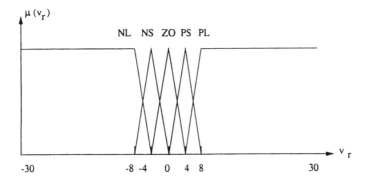

Fig. 1.21. Membership functions of v_r.

The universe of discourse of each variable depends in part on its physical range. However, the proper selection of the universe can improve the performance of the fuzzy system. Trial and error may be necessary in this process.

3. Extraction of fuzzy rules from expert knowledge and common sense. In this example, it is straightforward to get the following rules:

R_1: If d_n is *Very Small* and v_r is *Negative Small*, then a is *Negative Big*;
R_2: If d_n is *Very Small* and v_r is *Zero*, then a is *Negative Medium*;
R_3: If d_n is *OK* and v_r is *Zero*, then a is *Zero*;

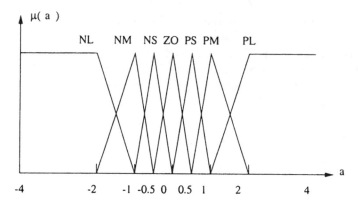

Fig. 1.22. Membership functions of a.

...

In this way, we can get a rule base as shown in Table 1.4.

Table 1.4. Fuzzy rule base for the distance controller

(d_n,v_r)	NB	NS	ZO	PS	PB
VS	NB	NM	NM	ZO	ZO
S	NB	NM	NS	ZO	ZO
OK	NB	NS	ZO	PS	PM
L	NM	ZO	PS	PM	PL
VL	NS	PS	PS	PM	PL

With the fuzzy rule base being generated and the membership functions being defined, a conclusion can be derived using approximate reasoning, for any given input. Fig. 1.23 is the input-output surface of the designed fuzzy controller.

1.4.2 Knowledge Representation with Fuzzy Preference Models

Besides IF-THEN rules, *a priori* knowledge can also be available that describes the relative importance of different variables or objectives, which are called preferences. Since human judgements, including preferences, are often imprecise, fuzzy logic also plays an important role in decision making and multi-criteria optimization [149].

Given n alternatives $o_1, o_2, ..., o_n$, preferences can be provided in the following three ways [34]:

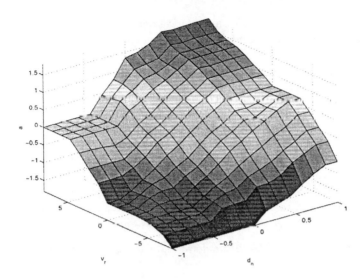

Fig. 1.23. Input-output surface of the obtained fuzzy controller.

- A preference ordering of the alternatives (ordered from the best to the worst):

$$O = \{o_1, o_2, ..., o_n\}, o_i \in [1, n]. \qquad (1.106)$$

For example, the ordering $O = \{2, 1, 3, 4\}$ means that alternative o_2 is the most important and o_4 is the least important.

- Utility values. In this case, a scaled real number is assigned to each alternative to indicate its relative importance:

$$U = \{u_i, i = 1, 2, ..., n\}. \qquad (1.107)$$

Utility values can be given in a difference scale or a ratio scale. If the ratio scale is used, we have $u_i \in [0, 1]$. If a difference scale is used, then u_i's are normalized so that

$$\max_i\{u_i\} - \min_i\{u_i\} \leq 1. \qquad (1.108)$$

- Fuzzy preference relations. In this case, the preferences are expressed by a binary relation matrix P of size $n \times n$, whose elements $p_{ij} \in [0, 1]$ is the preference x_i over x_j that satisfies the following conditions:

$$p_{ij} + p_{ji} = 1, \qquad (1.109)$$
$$p_{ii} = 0.5. \qquad (1.110)$$

Note that both the utility functions and the fuzzy preference relations mentioned above are numeric. Naturally, there are also utility functions and fuzzy preferences in linguistic forms:

- *Fuzzy utility functions*. Each alternative has a linguistic utility such as *very important, important, don't care, not important* and so on.
- *Linguistic fuzzy preference relations*. For example, instead of assigning a numeric fuzzy preference $p_{ij} = 0.8$, one can also express his judgement as "Objective x_i is *more important than* x_j".

Among the three preference models, fuzzy preference relations are the most widely used model. Several issues in fuzzy preference relations worth further discussion. Transitivity is one of the most important albeit disputing issue. Although there are examples in real world in which intransitivity of preferences occurs, it is still very helpful in applications to require the transitivity of preferences. Basically, transitivity means that if x_i is preferred to x_j and x_j is preferred to x_k, then x_i should be preferred to x_k. Some definitions of transitivity of fuzzy preference relations are given in the following.

- *Max-min transitivity* [50] or *moderate stochastic transitivity* [162], which is defined as

$$p_{ij} \geq 0.5, \ p_{jk} \geq 0.5 \Longrightarrow p_{ik} \geq \min(p_{ij}, p_{jk}), \ \forall i, j, k. \qquad (1.111)$$

- *Restricted max-max transitivity* [50] or *strong stochastic transitivity* [162]:

$$p_{ij} \geq 0.5, \ p_{jk} \geq 0.5 \Longrightarrow p_{ik} \geq \max(p_{ij}, p_{jk}), \ \forall i, j, k. \qquad (1.112)$$

- *Additive transitivity* [221], which is a stronger concept than the restricted max-max transitivity:

$$(p_{ij} - 0.5) + (p_{jk} - 0.5) = p_{ik} - 0.5, \forall i, j, k. \qquad (1.113)$$

- *Multiplicative transitivity*:

$$(1/p_{ij} - 1)(1/p_{jk} - 1) = 1/p_{ik} - 1, \forall i, j, k. \qquad (1.114)$$

Another important issue is the relationship between the preference models, namely, the fuzzy preference relation, the ordering and the utility functions. Consider a utility function $u(x)$ based on a difference scale, then the utility values can be transformed into the fuzzy preference relations as follows:

$$p_{ij} = \frac{1}{2}(1 + u_i - u_j). \qquad (1.115)$$

It can be proved that the fuzzy preference relations resulting from the above transformation satisfy the additive transitivity [221].

If the utility values are given on a ratio scale, then the following equation will convert the utility values into fuzzy preference relations that satisfy the multiplicative transitivity [221]:

$$p_{ij} = \frac{u_i}{u_i + u_j}. \qquad (1.116)$$

With the assumption that $u(x) \in [0, 1]$, a generic function can be defined to transform utility values to a fuzzy preference relation [34].

Similarly, a preference ordering can also be transformed into fuzzy preference relations. Given an ordering $O = \{o_1, o_2, ..., o_n\}$ arranged from the best to the worst, an example function to transform the ordering into fuzzy preference relations is

$$p_{ij} = \frac{1}{2}\left(1 + \frac{o_j - o_i}{n - 1}\right). \tag{1.117}$$

Preferences have found rich applications in multi-criteria decision making [57] and evolutionary multiobjective optimization [38], with the help of the fuzzy preference models. In this way, a priori knowledge can be used to solve various problems.

1.4.3 Fuzzy Group Decision Making

In practice, usually a group of experts make judgements and provide preference decisions. Suppose there are m experts and their preferences of alternative x_i over x_j are p_{ij}^l, where $i, j = 1, 2, ..., n$ and $l = 1, 2, ..., m$. Therefore, there are m preference matrices $P^l, l = 1, 2, ...m$ of size $n \times n$. In decision making, an alternative can be chosen either based on the *direct approach* or the *indirect approach* [131, 88]. In the direct approach, the solution is directly obtained from the individual preferences P^l. Three methods of indirect approach are often used, namely the *fuzzy α-core method*, the *alpha-minmax set method* [177] and the *mini-max degree set method* [131].

In contrast, the indirect approach derives a collective preference P^c from the individual preferences P^ls and then using the P^c to get a solution. Two different types of methods can be used depending on whether the preferences are numeric or linguistic. If the fuzzy preferences are numeric, a collective preference can be obtained by

$$p_{ij}^c = \frac{1}{m}\sum_{l=1}^{m} p_{ij}^l. \tag{1.118}$$

It is transitive if all the individual preferences are transitive. Alternatively, one can first define a binary preference a_{ij}:

$$a_{ij}^l = \begin{cases} 1 & \text{if } p_{ij}^l > 0.5 \\ 0.5 & \text{if } p_{ij}^l = 0.5 \\ 0 & \text{if } p_{ij}^l < 0.5 \end{cases}. \tag{1.119}$$

Then the collective preference is defined by

$$p_{ij}^c = \frac{1}{m}\sum_{l=1}^{m} a_{ij}^l. \tag{1.120}$$

Another type of collective preference relation can be obtained by

$$p_{ij}^c = \frac{\Pi_{l=1}^m p_{ij}^l}{\Pi_{l=1}^m p_{ij}^l + \Pi_{l=1}^m p_{ji}^l}. \tag{1.121}$$

However, a collective fuzzy preference relation derived using this method does not necessarily satisfy the transitivity condition.

The direct [227, 89] approaches to group decision making in the form of the linguistic fuzzy preference relations are generally based on two different preference degrees: a *linguistic non-dominance degree* defined in [179] and a *dominance degrees* using the concept of *fuzzy majority* [131, 87]. In [89], an indirect approach is proposed using the *linguistic ordered weighted aggregation* (LOWA) operator [246]. In non-fuzzy case, a majority measure can be either absolute or relative. A relative non-fuzzy majority can be achieved by taking the alternative that most of the group members have chosen. An absolute non-fuzzy majority is achieved when one alternative has been chosen by more than 50% of the group members. In the fuzzy case, a majority degree can be defined using *the fuzzy linguistic quantifiers* introduced in [255]. The fuzzy linguistic quantifiers are linguistic terms such as *most, at least half* and *as many as possible*. Fig.1.24 is an example of the definition for linguistic quantifiers *most, at least half and as many as possible* when the fuzzy linguistic quantifiers are defined by fuzzy sets in $[0, 1]$. A *relative non-decreasing linguistic quantifier* $Q(x)$ has the following properties:

$$Q(0) \quad = \quad 0, \tag{1.122}$$

$$\exists x_0 \ \text{ such that } \ Q(x_0) = 1, \tag{1.123}$$

$$Q(a) \geq Q(b) \quad \text{if} \quad a > b. \tag{1.124}$$

$$\tag{1.125}$$

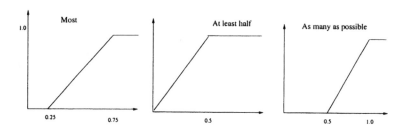

Fig. 1.24. An examples of linguistic quantifiers: Most, At least half and As many as possible.

Given m individual preferences $P^l, l = 1, 2, ..., m$, a linguistic quantifier and a weighting vector $W = (w_1, w_2, ..., w_m)$ that is used to represent the given fuzzy majority, a collective preference using the fuzzy majority based on LOWA can be obtained with the following procedure [89]:

1. Order the labels $\{P^l, l = 1, 2, ..., m\}$ so that $P^1 \geq P^2 \geq ... \geq P^m$. Here $\geq$ is used in a fuzzy sense, for example, *much more important*, *more important*.

2. Calculate the weights according to [246]

$$w_i - Q(i/m) \quad Q((i \quad 1)/m), i - 1, 2, \quad , m \quad (1.126)$$

The membership function $Q(x)$ is defined based on the given linguistic quantifier.

3. The collective preference is defined by:

$$C\{(P^1, P^2, ..., P^m), (w_1, w_2, ..., w_m)\} = w_i \odot P^1 \oplus (1 - w_1)C\{\alpha_h, P^h\}, \quad (1.127)$$

where, $h = 2, 3, ..., m$ and

$$\alpha_h = w_k / \sum_{k=h}^{m} w_k, h = 2, 3, ..., m, \quad (1.128)$$

which can be calculated in an iterative way:

$$C\{(P^1, P^2, ..., P^m)\} = C^2\{P^1, C^{m-1}(P^2, ..., P^m)\} \quad (1.129)$$
$$= C^2\{P^1, C^2\{C^{m-2}(P^3, ..., P^m)\}\} \quad (1.130)$$
$$= ..., \quad (1.131)$$

where $C^2\{(P^1, P^2), (w_1, w_2)\} = P^s(i \geq j)$ such that

$$s = \min\{T, j + \text{round}(w_i \cdot (i - j))\}. \quad (1.132)$$

In the above equation, T is the number of linguistic labels in total.

Fuzzy group decision methods provide a way of aggregating individual preferences of a group of members in order to obtain a single collective preference. However, the preferences are different and sometimes conflicting. As pointed out by *Arrow's Impossibility Theorem* [3], it is generally impossible to combine individual preferences into a collective preference without violating assumptions on individual sovereignty, non-dictatorship, and so on [136]. Of course, decision making processes that are absolutely democratic and individual-respecting do not exist in the real world.

2. Evolutionary Algorithms

2.1 Introduction

The basic idea behind *evolutionary algorithms* is to simulate the natural process of evolution to achieve systems that are able to improve themselves by applying three main *genetic operations* called *recombination, mutation* and *selection*. Through these conceptually simple operations, evolutionary algorithms are able to exhibit very attractive properties such as self-organization, adaptation, optimization and creation. Due to these properties, evolutionary algorithms have been widely applied in the fields ranging from the basic research on biological intelligence such as *artificial life* [151] to real word applications such as design optimization [181].

2.2 Generic Evolutionary Algorithms

The origin of evolutionary algorithms can be traced back to at least the 1950's [21, 68]. *Genetics algorithms* [92, 72], *evolution strategies* [198, 5], and *evolutionary programming* [60, 59] are three main approaches to evolutionary computation. *Genetic programming* [144, 150], which was developed on the basis of genetic algorithms, has been receiving increasing popularity in the recent years. A generic evolutionary algorithm can be described as follows:

$$
\begin{aligned}
&\text{begin EA;}\{\\
&t = 0;\\
&\text{generate initial population } p(0);\\
&\text{evaluate } p(0);\\
&\text{do } \{\\
&\quad t = t + 1;\\
&\quad \text{select parents from population } p(t).\\
&\quad \text{recombine } p(t);\\
&\quad \text{mutate } p(t);\\
&\quad \text{evaluate } p(t);\\
&\quad \} \text{ while } \{\text{stop criteria are not met }\};\\
&\}
\end{aligned}
$$

In both natural and artificial systems, there are three levels of evolution, i.e. *the genotype space, the phenotype space* and *the fitness space*. In evolutionary algorithms, a set of chromosome is called the *genotype*. Consequently, the genotype defines a *phenotype* with a certain *fitness*. Whereas the genotype is a set of chromosome, the definition of phenotype is problem-related. For example, if an evolutionary algorithm is used to evolve a fuzzy system, the phenotype consists of all fuzzy rule bases that may result from the genotype. If an evolutionary algorithm is used to optimize the structure of an artificial neural network, the phenotype includes all the possible structures of the neural network. The introduction of the phenotype may contribute to a sensible analysis of evolutionary algorithms, because the analysis of genotype-fitness mapping (also known as *fitness landscape*) may be more difficult. Recall that all genetic variations, such as recombination and mutation happen on the genotype space, which directly results in variations in the phenotype space.

In practice, the fitness space is defined by the given task. For instance, for an evolutionary fuzzy system, its fitness reflects the modeling error if the fuzzy system is used to model an known system. However, if the fuzzy system is used as a controller, the fitness function is dependent on the control error. In other words, the same fuzzy system has different phenotype-fitness mappings for different purposes.

Although the three main evolutionary algorithms have great similarity in structure, they differ in several aspects, including representation, population sizing, recombination, mutation and selection. Since both genetic algorithms and evolution strategies are employed in this book, the basic operations of these two algorithms will be introduced.

2.2.1 Representation

Traditionally, genetic algorithms use the *binary* or gray coding [30], in which object parameters [1] are encoded into a string consisting of 0's and 1's. Suppose the object parameters are $\mathbf{x} = [x_1, x_2, ..., x_n]$, then the chromosome looks like

$$\underbrace{101...1}_{x_1}\underbrace{110...0}_{x_2}...\underbrace{111...1}_{x_n}. \qquad (2.1)$$

If the binary coding is adopted and l bits binary numbers $s_j, j = 1, 2, ..., l$ are used to represent a real number x_i that is within the range of $[a_i, b_i]$, then the decoding function will be:

$$x_i = a_i + (b_i - a_i)\frac{1}{2^l - 1}\left(\sum_{j=0}^{l-1} s_j 2^j\right). \qquad (2.2)$$

---

[1] We distinguish between object parameters and strategy parameters. Object parameters are the variables of the fitness function to be optimized, whereas the strategy parameters are the parameters of the evolutionary algorithm, such as the population size, the recombination probability, the mutation step size and so on.

It is noticed that the binary coding is non-linear, which is undesirable when the evolutionary algorithm approaches to the optimum. As an alternative, the gray coding can also be used. The most important feature of the gray coding is that the decimal value increases or decreases by 1 if only one bit is changed. In decoding, the gray-coded string can first be converted into a binary string and then equation (2.2) can be used.

In the above coding schemes, the length of chromosome is fixed. However, it is known that biological genomes are of variable lengths, are independent of position and may contain duplicative or competing genes [184]. An early work on variable length representation was the *messy GA* [73], in which both the value and position of each bit are encoded in the chromosome. The delta-coding was suggested in [165] that uses variable length representations to change the range of coding so that the genetic algorithm is able to get a good balance between exploration and exploitation.

The binary and gray coding have been working successfully for parameter optimization problems. However, for more complicated tasks, such as the structure optimization of neural networks, more sophisticated coding schemes are necessary. Examples are grammar coding [138], cellular coding [80], recursive coding [203] and context dependent coding [156] among others. These coding methods are sometimes called indirect coding because not every property in the phenotype has explicit representation in the genotype. For example, the context dependent coding (CDC) has two types of genes, one for instructions and the other for numbers. The instruction are used to represent parameters to be optimized. Therefore, a chromosome may have some genes that will be completely ignored in decoding. Suppose A and B are used for instruction coding, and 1 and 0 are used for number coding, then a typical chromosome can look like as follows:

$$A0BA1011AAB101B00B1B.$$

In decoding, we first look for an instruction, which will be coded as AA, AB, BA, or BB. Therefore, the first two genes are ignored until BA are found. The numeric genes are then looked for. In this case, 1011 are the numeric genes to be used by instruction BA. The reading of the numeric genes terminates until the next instruction gene is encountered. Since numeric genes will be expected after the instruction genes AA, B is ignored.

Generally, indirect coding is used for structure optimization of neural networks [201] and fuzzy systems [156]. Variations in both length and structure of the coding have led to new evolutionary algorithm - genetic programming [144].

In contrast, evolution strategies encode the object parameters directly in the chromosome. Therefore, no coding and decoding is needed in evolution strategies. However, one important feature in the representation of evolution strategies is that both the strategy parameters and the object parameters of the algorithm are encoded in the chromosome. Therefore, a chromosome for an object vector $\mathbf{x} = [x_1, x_2, ..., x_n]$ may look like

$$x_1 x_2 ... x_n \mathbf{p_1 p_2 ... p_n}, \qquad\qquad (2.3)$$

where $\mathbf{p_i}, i = 1, 2, ..., n$ are the vector of the strategy parameters for object parameter x_i. The number of strategy parameters for each object parameter is different in various evolution strategy algorithms.

It is seen that genetic algorithms have more flexible representations than evolution strategies, which may be one reason why genetic algorithms have found applications in a much wider range of fields. On the other hand, different representations exhibit different *causality* properties between the genotype and phenotype of an evolutionary algorithm. By causality, it is meant that the degree of variations in the genotype space should be properly reflected by the degree of variation in the phenotype space. That is to say, the neighborhood in the genotype space should be preserved in the phenotype space [202]. As found in [202], causality is not only a very important condition for the *self-adaptation* in evolutionary algorithms, it is also necessary at the exploitation stage of evolutionary algorithms. By defining a quantitative causality measure, it was found that the representation of evolution strategies has stronger causality than the binary coding used in genetic algorithms [202]. Further, it was also shown that gray coding is of stronger causality than binary coding. This may be an explanation for why evolution strategies have exhibited better performance than genetic algorithms in parameter optimization problems that we have experienced.

Meanwhile, it is well known that biological chromosomes are highly redundant [39]. Research in artificial evolutionary systems has also shown that redundant representation is less *epistatic* and makes the evolutionary algorithms more robust, particularly in changing environments [44, 81]. In genetic algorithms, epistasis is defined as the influence of the genotype at one locus on the effect of a mutation at another locus [232].

Alternatively, the redundancy in the genotype-phnotype mapping has also been investigated in terms of fitness landscape *neutrality*. By fitness landscape neutrality, it is meant that the fitness value keeps the same despite the movement in the genotype space. Neutral mutation will occur when the population moves between solutions of equal fitness, which are termed *neutral networks*. This kind of neutrality is believed to be helpful in escaping from local optima. In addition, neutrality also plays a positive role in adaptation [100].

One new form of representation that has attracted attention is call *embryogeny*. By embryogeny, the genotype is regarded as a set of growing instructions that define how the phenotype will develop [14]. Some of the indirect coding schemes in genetic algorithms, such as the cellular encoding used in [145] and the Lindenmayer systems used in [36] can also be regarded as embroyogenies.

2.2.2 Recombination

Recombination (crossover) provides a mechanism for information exchange between different individuals. It was believed to be the most powerful genetic operator in genetic algorithms. Although Holland was not the first to suggest recombination in evolutionary algorithms [65], he emphasized this operator in his schemata theory [92]. According to Holland, an adaptive system must be able to find new structures that have a high probability of improving performance significantly. Schemata, as defined by Holland, provide a basis for associating combinations of attributes with potentially better performance. Recombination is often referred as crossover and we do not distinguish between the two notations.

Many methods have been proposed to implement recombination to improve the performance of the genetic algorithms for different problems. Actually, the method of recombination depends on the method of representation, i.e. the genetic coding. Generally, recombination can be divided into genotype-level schemes and phenotypic-level schemes [209]. For example, genetic algorithms that using binary or other bit-based coding schemes use genotypic-level schemes. The most popular genotypic-level recombination methods are:

- *1-point recombination* [92]. Suppose there are two parents for recombination: $[AB1|01A001]$ and $[BA0|10A110]$. One point (denoted by '—') on the chromosome is randomly selected, say the third bit. Two offspring are generated by swapping two segments cut by the selected point: $[BA001A001]$ and $[AB110A110]$.
- *n-point recombination* [45]. 1-point recombination was extended to n-point recombination. In this scheme, n cut-points are introduced. For example, we suppose the two-point crossover is implemented on the two parents that are cut at the third and sixth loci: $[AB1|01A|001]$ and $[BA0|10A|110]$. Thus, the following two offspring are created: $[AB110A001]$ and $[BA001A110]$.

  Multi-point crossover can be also extended to two or n-dimensional schemes [133]. A chromosome can be tessellated into $i \times j$ blocks, then genes in a block are swapped between two parents. More generally, the chromosome can be broken into complementary fragments. Then different operators can be defined to process these fragments for generating offspring.
- *Uniform recombination* [216]. Uniform recombination does not use cut-points. Instead, it decides bit by bit whether the particular gene should be swapped between the two parents. Usually, the probability of swapping is set to 0.5, although other probability can also be used. If the swapping probability is 0.5, the uniform recombination can be complemented in the following way. First, a binary decision string with 0 and 1 is generated randomly. The length of the decision string is the same as the parents. If the i-th bit of the decision string is 1, then swap the i-th bit of the parent

chromosome. For example, if the generated decision string is [011010010]. Thus the following two offspring [$AA000A011$] and [$BB111A100$] will be generated from the two patents [$AB101A001$] and [$BA010A110$]. It is believed that the uniform recombination is more disruptive of schemata than 1-point and 2-point recombination, while it is not length-bias. It is noticed that a highly disruptive recombination can be both positive and negative, because it is able to introduce more new schemata, however, it could destroy an existing good schemata.

Phenotypic-level recombinations are often used in evolution strategies. In [198], the following four types of recombination were implemented:

$$
x_i' = \begin{cases}
x_{a,i} \text{ or } x_{b,i}, & \text{discrete recombination} \\
\frac{1}{2}(x_{a,i} + x_{b,i}), & \text{intermediate recombination} \\
x_{a,i} \text{ or } x_{b_j,i}, & \text{global discrete recombination} \\
\frac{1}{2}(x_{a,i} + x_{b_j,i}), & \text{global intermediate recombination}
\end{cases}
\tag{2.4}
$$

where a, b, b_j are parent individuals and x_i is the i-th element of vector x. Note that in the global recombination methods, three individuals are involved. This can of course be extended to several individuals, which is called the *panmictic recombination* [5]. In addition, the weighting for each individual can also be parameterized.

2.2.3 Mutation

Mutation was regarded as a secondary operator to crossover in genetic algorithms. The main purpose of mutation is to prevent *premature convergence*, when all individuals in the population become very similar before the GA converges to the global optimum. When all the chromosomes are very similar or even the same in a population, it is hard to create a new structure (schema) only by recombination. Mutation simply makes an inversion of a bit if the binary or the gray coding schemes are used.

Contrary to the genetic algorithms, mutation is the main operator in evolution strategies. In evolution strategies, mutation is done by adding a normally distributed random number to the object parameters to be mutated.

The relative importance of recombination and mutation is one of the major issues in genetic algorithms. Usually, recombination is believed to be more powerful and mutation alone is insufficient. In fact, the power of a genetic operator can be evaluated in two aspects: *exploitation* and *exploration*, or as sometimes called, *construction* and *disruption*. Exploitation is the ability to approach an optimum gradually, whereas exploration is the ability to create new optima. Most of the comparisons are based on empirical studies, and therefore, different or even contradictory conclusions have been drawn. However, a general conclusion is that recombination plays a more important role than mutation if the population size is large. On the contrary, mutation is more important if the population size is small [170].

2.2.4 Selection

Selection embodies the natural law of *survival of the fittest*. In evolutionary algorithms, the main concern in choosing a selection method is to maintain a proper selection pressure. The often used selection schemes for genetic algorithms are the *Roulette wheel selection* [72], the *ranking selection* and the *tournament selection* [74]. All the above selection schemes are stochastic, that is to say, the best individuals may be lost although their probability of being selected is high. To address this problem, the *elitism strategy* can be introduced. In an elitism strategy, the best individual(s) are always chosen for the next population so that the best individual ever seen will not be lost. It has been shown that elitism plays an important role in the convergence of GA [191]. More detailed discussion on selection methods can be found in [146].

There are two selection schemes in evolution strategies, namely (μ, λ)-selection and $(\mu + \lambda)$-selection, where μ, λ are the population size for the parents and the offspring. In (μ, λ)-selection, the best μ individuals in the offspring are chosen as the parents for the next generation, whereas in $(\mu+\lambda)$-selection, the best μ individuals from the union of parents and offspring are chosen as the parent individual for the next generation. $(\mu + \lambda)$ can be seen as a method with μ elitists. Note that both of the selection methods in evolution strategies are deterministic.

2.3 Adaptation and Self-Adaptation in Evolutionary Algorithms

Adaptation and self-adaptation in evolutionary algorithms mean the modification of the parameters of the evolutionary algorithms, usually known as the strategy parameters, during operation based on either heuristics or the feedback of the environment. The parameters of evolutionary algorithms include any elements that parameterize the evolutionary algorithm, including the population size, the chromosome length, the coding scheme, the recombination and mutation probability, the mutation step-size and so on. If the modification is carried out by a separate mechanism from the evolution process, it is usually called *parameter adaptation*. In contrast, if the modification is implemented by encoding the strategy parameters in the chromosome, it is usually termed *self-adaptation*.

2.3.1 Adaptation

Adaptation of the strategy parameters are most often employed in genetic algorithms. For example, the adaptation of the population size has been introduced in [72] and a dynamic parameter encoding method has been suggested in [197].

Besides the static recombination methods that have been introduced previously, several adaptive recombination method have also been proposed. One idea is to identify groups of genes that should be kept together when crossover occurs [241]. That is to say, it is possible to learn from the previous crossover operations in order to minimize the possibility of destroying the highly fit schemata. This can be done by using some reward and penalty rules. For example, the following rules can be generated:

If offspring fitness is higher than parent's fitness, then reward bits which come from parents with probability p_{Reward}.

Another way of choosing good recombination is to label a crossover with *good*, *bad* or *inactive* [200]. A crossover that is inactive means that the fitness of offspring is the same as that of the parent using the particular crossover.

In a search process, it is very important to balance the exploration and the exploitation at different search stages. Heuristically, the algorithm should focus more on exploration in the beginning and more on exploitation at the end of the search. According to this heuristics, several methods have been developed to adapt the recombination and mutation probability [58]. For example, the mutation rate is set to a relatively high value and is gradually reduced during the evolution:

$$p_m(t) = 0.5 - 0.3 \cdot \frac{t}{T}, \tag{2.5}$$

where, t is the generation index, T is overall number of generations to be run in the evolution. Alternatively, the mutation rate can be modified as follows:

$$p_m(t) = \alpha \cdot \frac{\exp\left(-\frac{\gamma t}{2}\right)}{P\sqrt{L}}, \tag{2.6}$$

where P is the population size, L is the length of the chromosome.

In evolution strategies, Rechenberg introduced the *1/5 success rule* to adapt the mutation step-size. If the success ratio is larger than $1/5$, then the step-size is increased. If the success ratio is smaller than $1/5$, the step-size is then reduced.

2.3.2 Self-adaptation

The adaptation of the mutation distribution is one of the the main issues in evolution strategies. Since in evolution strategies, the strategy parameters are evolved together with the object parameters, the modification of the strategy parameters are called the self-adaptation. Several methods have been developed, such as the standard mutation [198], the *rotation matrix adaptation* method that controls not only the step size but also the direction of the mutation [198], the *generating set adaptation* approach which adapts normal mutation distributions independent of the chosen coordinate system [84], and the de-randomized *covariance matrix adaptation* [180].

In the following, the standard adaptation and the covariance matrix adaptation method will be presented briefly.

Standard Adaptation. In the standard evolution strategies, mutation is carried out in the following way:

$$x_i(t) = x_i(t-1) + \sigma_i(t-1) \cdot N(0,1), \tag{2.7}$$

$$\sigma_i(t) = \sigma_i(t-1) \cdot \exp(\tau' \cdot N(0,1) + \tau \cdot N_i(0,1)), \tag{2.8}$$

where, i is the index for the i-th object parameter, τ and τ' are constants, $N(0,1)$ and $N_i(0,1)$ are zero-mean normal distributions.

Covariance Matrix Adaptation. The mutation algorithm in the standard evolution strategies can be rewritten in a vector form:

$$\boldsymbol{x}(t) = \boldsymbol{x}(t-1) + \boldsymbol{\sigma}(t-1)N(\boldsymbol{0}, \boldsymbol{I}), \tag{2.9}$$

where $N(\boldsymbol{0}, \boldsymbol{I})$ is the normal distribution known as the mutation distribution. A more general mutation distribution can be used as follows:

$$\boldsymbol{x}(t) = \boldsymbol{x}(t-1) + \boldsymbol{\sigma}(t-1)N(\boldsymbol{0}, C(t-1)), \tag{2.10}$$

where $\boldsymbol{\sigma}(t-1)$ is the global step-size in generation $t-1$, $N(\boldsymbol{0}, C(t-1))$ is the normal distribution with a covariance matrix of $C(t-1)$. Let $C = \boldsymbol{B}\boldsymbol{B}^T$, then the modification of the object parameters can be written as:

$$\boldsymbol{x}(t) = \boldsymbol{x}(t-1) + \sigma(t-1)\boldsymbol{B}(t-1)\boldsymbol{z}, \; z_i \sim N(0,1). \tag{2.11}$$

The adaptation of the covariance matrix is carried out in the following way:

$$C(t) = (1 - c_{\text{cov}})C(t-1), + c_{\text{cov}}\boldsymbol{s}(t)\boldsymbol{s}^T(t) \tag{2.12}$$

$$\boldsymbol{s}(t) = (1-c)\boldsymbol{s}(t-1) + \sqrt{c(2-c)}\boldsymbol{B}(t-1)\boldsymbol{z}, \tag{2.13}$$

where c is a constant that is heuristically set to $1/\sqrt{n}$, c_{cov} is a constant that is heuristically set to $2/n^2$, $\boldsymbol{s}$ is called the evolution path [83]. The global step-size is modified as follows:

$$\sigma(t) = \sigma(t-1)\exp(\beta(\|\boldsymbol{s}_\sigma(t)\| - \hat{\chi}_n)), \tag{2.14}$$

$$\boldsymbol{s}_\sigma(t) = (1-c)\boldsymbol{s}_\sigma(t-1) - \sqrt{c(2-c)}c_w\boldsymbol{B}(t-1)\boldsymbol{z}, \tag{2.15}$$

where $\hat{\chi}_n = E[\|N(\boldsymbol{0}, \boldsymbol{I})\|]$ is the expected length of a normally distributed vector that is approximately equal to $\sqrt{n}(1 - \frac{1}{4n} + \frac{1}{21n^2})$, c_w is a normalization parameter so that $\boldsymbol{B}(t-1)\boldsymbol{z}$ is $N(\boldsymbol{0}, \boldsymbol{I})$ distributed.

The main idea behind the covariance matrix adaptation is to change the mutation distribution so that the probability of producing a successful mutation step again is increased. A mutation step is considered to be successful if the individual that carries out this mutation step is selected in reproduction.

2.4 Constraints Handling

Constraints in optimization are very common and therefore handling constraints properly is an important topic in genetic algorithms. Usually, linear constrains include equality and inequality constraints. For example, to maximize a function $f(x_1, x_2, ..., x_n)$, there usually exist the following constraints [166]:

- *Domain constraints*: $a_i \leq x_i \leq b_i$ for $i = 1, 2, ..., n$, a_i and b_i are the lower and upper bounds of the parameter. For instance, in optimizing the parameters of the Gaussian fuzzy membership functions, there will be domain constraints for the center value of the Gaussian function. There are also other domain constraints. Suppose we want to select three sensors from a number of sensors that are labeled with an identification number. At the same time, the sensors should be able to detect a particular target. In Fig. 2.1, only 10 sensors are able to detect the target. Therefore, only these sensors are legal.
- *Equalities*: $Ax = b$, where $x = (x_1, ..., x_n)$, $A = (a_{ij})$, $I = 1, 2, ..., n; j = 1, 2, ..., m$, and m is the number of equations, $b = (b_1, ..., b_m)$.
- *Inequalities*: $Cx \leq d$, where $x = (x_1, ..., x_n)$, $C = (C_{ij})$, $d = (d_1, ..., d_l)$, where $i = 1, 2, ..., n; j = 1, 2, ..., l$, and l is the number of inequalities.

A number of approaches to dealing with constraints have been proposed. The domain constraints given in the form of the upper and lower bound are easy to handle. It can be seen from the binary or gray encoding and decoding schemes introduced in the previous section that such domain constraints can be directly satisfied in problem representation. For the domain constraints in other forms, specific methods should be developed. For equality and inequality constraints, the following approaches can be adopted:

- *Penalty function* [208]. Equalities and inequalities could be treated as a part of the penalty function, which will play a role if the constraints are violated. However, the penalty function may make the genetic algorithm very inefficient if most of the individuals evaluated do not satisfy the constraints.
- *Multiple objective optimization* [213, 37]. An alternative to use a penalty function is to treat a constraint as an objective of the evolutionary algorithm. This approach becomes complicated when the number of constrains is high.
- *Repair algorithms* [166]. It is possible to correct the infeasible solutions in order to satisfy the constraints. We provide here a simple example. Consider the following equality constraint in optimizing $f(x_1, x_2)$:

$$x_1 + x_2 = 1. \tag{2.16}$$

Due to mutation, x_1 is changed to $x_1 + c$. To satisfy the constraint, x_2 is modified in the following way:

$$x_2' = x_2 - c, \tag{2.17}$$

assuming that $x_2 - c$ is still within the domain of x_2. It can be seen that repair algorithm is very problem-specific and will become very complicated for complex problems.

- *Incorporating a priori knowledge.* A *priori* knowledge about the constraints can be incorporated into genetic operators so that only valid solutions will be generated in the evolution [185].

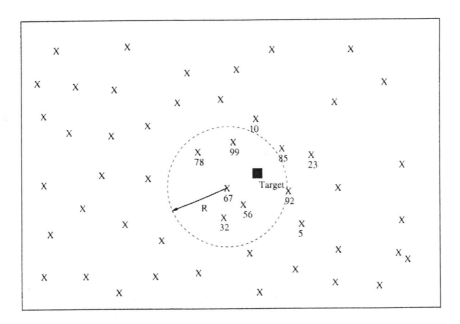

Fig. 2.1. Sensor self-organization for target tracking. Only 10 sensors that can detect the target and are labeled. In the figure, 'X' denotes an acoustic sensor.

Ad hoc methods can also be developed for specific problems in dealing with constraints. Take the sensor self-organization for target tracking as an example [26]. Suppose there are 50 sensors distributed randomly in a given area, refer to Fig.2.1, where an 'X' represents an acoustic sensor. The detection range of a sensor is a circle of radius R, therefore, only part of the given sensors can detect a target moving in the area. For bearing-only sensors, at least three sensors are needed to locate the position of the target. Since it is not desirable to turn on all the sensors all the time, genetic algorithms can be used to select the sensors that need to be switched on to maximize the tracking accuracy and to minimize the energy consumption of the sensors. Remember that only a small part of the 50 sensors can see the target. To prevent the genetic algorithm from generating invalid solutions, the sensors that can detect the target should be re-labeled. Assume 10 sensors can detect the target with original labels listed in the first row of Table 2.1, then they can be mapped to ten consecutive integers before they are encoded in the chromosome.

Table 2.1. Re-label of sensors to remove the constraints

Original label of sensors	67	56	32	23	5	78	99	85	92	10
New label of sensors	0	1	2	3	4	5	6	7	8	9

2.5 Multi-objective Evolution

Most real-world optimization problems have more than one objective and the objectives are usually conflicting. In this case, a solution is said to be *Pareto optimal*. Before introducing the definition of Pareto optimality, we first need to give the concept of *Pareto dominance*:

Pareto Dominance: A vector $\mathbf{u} = (u_1, ..., u_n)$ is said to dominate another vector $\mathbf{v} = (v_1, ..., v_n)$ if and only if $\forall i \in \{1, ...n\}$, $u_i \leq v_i$ and $\exists i \in \{1, ..., n\}$, so that $u_i < v_i$ (for minimization problems).

Based on the definition of the Pareto dominance, the *Pareto optimality* can be defined as follows:

Pareto Optimality: A solution $x^\star$ is said to be Pareto optimal if and only if there is no x for which $\mathbf{v} = \mathbf{F(x)} = (f_1(\mathbf{x}), ..., f_n(\mathbf{x}))$ dominates $\mathbf{u} = \mathbf{F(x^\star)} = (f_1(\mathbf{x^\star}), ..., f_n(\mathbf{x^\star}))$, where $\mathbf{F(x)}$ is an n-dimensional vector of objective functions.

Generally, the entire collection of Pareto optimal solutions are called the Pareto optimal sets and the corresponding set of the objective vectors ($\mathcal{P}$) is denoted as the *Pareto-optimal front*, for short, the Pareto front. The Pareto front is said to be *convex* if and only if $\forall \mathbf{u}, \mathbf{v} \in \mathcal{P}, \forall \gamma \in (0, 1), \exists \mathbf{w} \in \mathcal{P}$: $||\mathbf{w}|| \leq \gamma ||\mathbf{u}|| + (1 - \gamma)||\mathbf{v}||$. Otherwise, the Pareto front is said to be *concave*.

A Pareto front can be neither purely convex nor purely concave, i.e., it is possible that part of the Pareto front is concave and part of it is convex, see for example in Fig. 2.2. In this case, the Pareto front is usually known as non-convex.

Traditional multi-objective optimization has been widely investigated in operation research (OR) [101]. Since the early work in [195], there has also been increasing interest in evolutionary multi-objective optimization (EMOO), refer to [238].

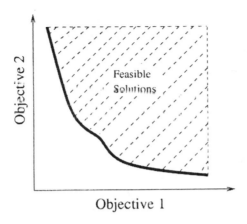

Fig. 2.2. An example of Pareto front. In the figure, the shaded area denotes all feasible solutions and the thick curve represents the Pareto front. This Pareto front is convex at both ends and concave in the middle.

In the following, the three main approaches to evolutionary multi-objective optimization are described briefly.

2.5.1 Weighted Aggregation Approaches

In this approach, the different objectives are summed up to one single scalar function. The advantage of this approach is that the algorithm is easy to implement and computationally inexpensive. All existing evolutionary algorithms can directly be applied with minor changes. The disadvantage of this approach is that *a priori* knowledge is needed to determine the weight for each objective and only one Pareto solution can be obtained from one run of optimization. One method for getting a set of Pareto-optimal solutions in one run of optimization has been suggested in [82]. To this end, the weights are encoded into the chromosome so that the genetic search is carried out in a diversified directions. In [117], two methods for changing weights during optimization are introduced. The first method is to generate a uniformly distributed weights in the population. Therefore, each individual searches a different direction. Furthermore, the weights are re-generated in each generation. Similar methods have been suggested in [103], in which random weights are assigned to each objective during selection. In that method, the number of search direction equals the number of parents that are selected for reproduction. The second method proposed in [117] is to change the weights gradually and periodically during the optimization from generation to generation. This is a very straightforward method and works effectively for both convex and concave Pareto-optimal problems. Theoretic analysis has been developed in [120] to explain the reason why this method works. Based on this theory, a surprisingly easy method has been suggested for obtaining a set of Pareto-optimal solutions, that is, to switch the weights between 0 and 1. This mechanism is very efficient if the Pareto front is concave. Chapter 9 will discuss the algorithms in more details.

2.5.2 Population-based Non-Pareto Approaches

The basic idea is to maintain the same number of sub-population as the objective number [195]. In each sub-population, the individuals are ranked and selected based on one single objective. Then, the offspring individuals selected from two sub-populations are merged and genetic operations (crossover and mutation) are performed, refer to Fig. 2.3 for bi-objective optimization. In this way, the whole population is able to approach to the Pareto front if certain diversity measure is taken. The main shortcoming of this algorithm is that the achieved Pareto solutions are only near-optimal. An alternative method is to choose one of the objective functions randomly during evaluation [148], see Fig. 2.4 for the two-objective optimization.

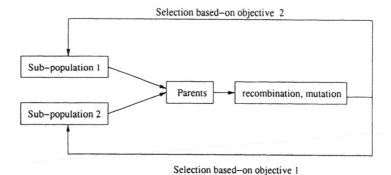

Fig. 2.3. The population-based non-Pareto approach.

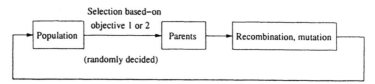

Fig. 2.4. Choose one of the objective function randomly to achieve multiobjective optimization.

2.5.3 Pareto-based Approaches

This is now the most popular approach to evolutionary multi-objective optimization [63, 94, 210]. Most of the Pareto-based approaches are developed from the original idea described in [72]. All the non-dominated individuals are assigned the same fitness value and others a inferior fitness. For example, in Fig. 2.5, no solution dominates solution A, therefore, the highest fitness value 1 is assigned to the solution (for minimization). Solution B is dominated by solution C, and thus the fitness value 2 is assigned to it. Since solution C is dominated by solution A and B, it gets the lowest fitness value of 3. In this approach, it is essential to take measures to maintain the diversity of the population. Besides, the performance can be improved significantly to intro duce elitism into the algorithm [258]. The main weakness of the approach is that the algorithms are computationally very expensive.

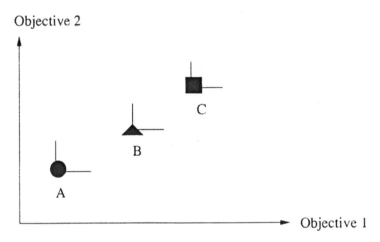

Fig. 2.5. The Pareto-based approach.

2.5.4 Discussions

Multiple objective optimization is receiving increasing interest in the evolutionary computation research community. Several issues are emerging to be important in evolutionary multiobjective optimization. Some of them are:

- Metrics. A large number of evolutionary multi-objective algorithms have been developed. However, it is generally difficult to compare the performance of different algorithms. One criterion has also been suggested in [63], which introduces the *attainment surfaces* to allow univariate statistical analyses on different Pareto sets. Using this criterion, one can know that one algorithm is better than the other, however, it is not straightforward to know how much better. Another criterion proposed in [258] takes the distance, the distribution and the extent of the solutions in the Pareto set into account. As pointed out in [139], wrong conclusion may be reached in some situations.

However, it is not enough to take into account the quality of the obtained solutions only in comparing the performance of the algorithms. It is equally important to consider the complexity of the algorithm, especially the number of fitness evaluations needed to achieve the set of solutions.

- Test functions. To investigate the performance of different algorithms, it is important to set up a number of test problems that are well understood and of representative importance of real-world problems. Issues that have been discussed so far are the multi-modality, deceptiveness, and nonlinear constrains that may make the problems hard for evolutionary algorithms. However, global convexity, connectedness of the Pareto-optimal solutions and the parameter-fitness mapping have not yet been well discussed in constructing test functions.

2.6 Evolution with Uncertain Fitness Functions

In real-world applications, there is always a degree of uncertainty in the fitness function. For example, to evaluate the performance of a turbine blade, it is necessary to solve a large number of partially differential equations. In solving the equations, numerical errors may be introduced. In this case, the fitness function is noisy. Other uncertainties occurs in case reduced models or approximate models are used in fitness evaluation to reduce the computation time.

2.6.1 Noisy Fitness Functions

The environment of the evolutionary algorithm may be noisy. Usually, there are three methods to deal with noisy fitness functions. It was shown that a larger population size helps if the fitness function is noisy. The second approach is to sample the fitness function several times and to average [55]. However, this method requires a large number of additional fitness evaluations. The third method is to calculate the fitness of an individual by averaging the value of this individual as well as that of other individuals in its neighborhood. To avoid additional computational cost, the individuals that participate in the averaging can be chosen from the current and previous generations [22]. A more flexible alternative is to estimate the fitness of the individuals in the neighborhood using a statistical model constructed with history data [193, 23].

2.6.2 Approximate Fitness Functions

So far, approximation of the fitness function in evolutionary computation has been applied mainly in the following cases.

- The computation of the fitness is extremely time-consuming. One good example is structural design optimization [122]. In aerodynamic design optimization, it is often necessary to carry out computational fluid dynamics (CFD) simulations to evaluate the performance of a given structure. A CFD simulation is usually computationally expensive, especially if the simulation is 3 dimensional, which takes over ten hours on a high-performance computer for one calculation of the fitness function. Therefore, approximate models have widely been used in structure optimization [8].
  Fitness approximation has also been reported in protein structure prediction using evolutionary algorithms [175]. A neural network has been used for feature extraction from amino acid sequence in evolutionary protein design.
- There is no explicit mathematical model for fitness evaluation. In many situations, such as in art design and music composition as well as in some areas of industrial design, the evaluation of the fitness depends on the human user. Generally, these problems can be addressed using interactive evolutionary computation [218]. However, a human user can easily get tired and an approximate model that embodies the opinions of the human evaluator is very helpful.
- The fitness landscape is multi-modal. The basic assumption is that a global model can be constructed to approximate and smooth out the local optima of the original multi-modal fitness function without changing the global optimum and its location [157]. Similar ideas have also been reported in conventional optimization methods. However, it is generally difficult to build an approximate model that has the same global optimum on the same location when the dimensionality is high with limited number of samples.
- The environment of the evolutionary algorithm is noisy. As previously discussed, one of the method to deal with noisy fitness function is to build an approximate model for the fitness function.

The application of approximation models to evolutionary computation is not as straightforward as one may expect. There are two major concerns in using approximate models for fitness evaluation. First, it should be ensured that the evolutionary algorithm converges to the global optimum or a near-optimum of the original fitness function. Second, the computational cost should be reduced as much as possible. One essential point is that it is very difficult to construct an approximate model that is globally correct due to the high dimensionality, ill distribution and limited number of training samples. It is found that if an approximate model is used for fitness evaluation, it is very likely that the evolutionary algorithm will converge to a false optimum. A false optimum is an optimum of the approximate model, which is not one of the original fitness function, refer to Fig. 2.6 for an example. Therefore, it is very essential in most cases that the approximate model should be used together with the original fitness function. This can be regarded as the issue of model management or evolution control. By evolution control, it is meant

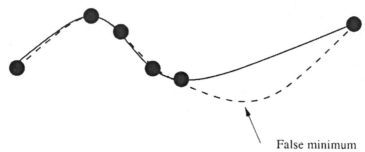

Fig. 2.6. An example of a false minimum in the approximate model. The solid line denotes the original fitness function, the dashed line the approximate model and the dots the available samples.

that in evolutionary computation using approximate models, the original fitness function is used to evaluate some of the individuals or all individuals in some generations [119]. An individual that is evaluated using the original fitness function is called a controlled individual. Similarly, a generation in which all its individual are evaluated using the original fitness function is called a controlled generation.

Generally, existing work on approximation in evolutionary computation can be divided into three main approaches from the viewpoint of evolution control.

- No Evolution Control. Very often, the approximate model is assumed to be of high-fidelity and therefore, the original fitness function is not at all used in evolutionary computation, such as in [17, 188, 130].
- Fixed Evolution Control. The importance to use both the approximate model and the original function for fitness evaluation has been recognized [186]. There are generally two approaches to evolution control, one is individual-based [78, 27], and the other is generation-based [186, 187]. By individual-based control, it is meant that in each generation, some of the individuals use the approximate model for fitness evaluation and others the original function for fitness evaluation. In individual-based evolution control, either a random strategy or a best strategy can be used to select the individuals to be controlled [119]. In the best strategy, the best individual (based on the ranking evaluated by the approximate model) in the current generation is reevaluated using the original function [78], see Fig. 2.7. To reduce the computational cost further, individual-based evolution control can be carried out only in a selected number of generations [27]. In contrast, the random strategy selects certain number of individuals randomly for reevaluation using the original fitness function [119]. An alternative to the best strategy and the random strategy is to evaluate the mean of the individuals in the current population [172].

Generation-based evolution control can also be implemented [186, 187]. In [186], generation-based evolution control is carried out when the evolutionary algorithm converges on the approximate model. More heuristically, evolution control is carried out once in a fixed number of generations, see Fig. 2.8.

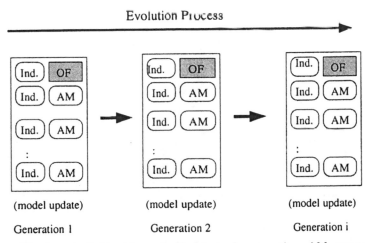

Fig. 2.7. The best individual is controlled in each generation. AM: approximate model; OF: original function.

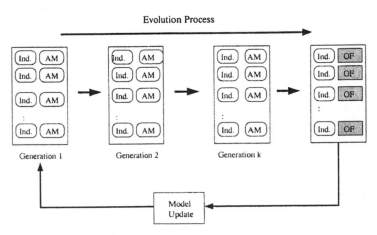

Fig. 2.8. Generation-based evolution control. AM: approximate model; OF: original function.

One drawback in the aforementioned methods is that the frequency of evolution control is fixed. This is not very practical because the fidelity of the approximate model may vary significantly during optimization. In fact, a predefined evolution control frequency may cause strong oscillation during optimization due to large model errors, as observed in [186].

- Adaptive Evolution Control. It is straightforward to imagine that the frequency of evolution control should depend on the fidelity of the approximate model. A method to adjust the frequency of evolution control based on the trust region framework [48] has been suggested in [172], in which the generation-based approach is used. A framework for approximate model management has also been suggested in [121], which has successfully been applied to 2-dimensional aerodynamic design optimization, see Fig. 2.9.

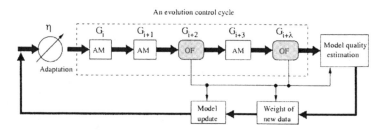

Fig. 2.9. Adaptive generation-based evolution control. In the evolution control cycle, there are λ generations, η $(\eta \leq \lambda)$ generations will be controlled. AM: approximate model; OF: original function.

2.6.3 Robustness Considerations

One issue in evolutionary optimization that is related to noisy and approximate fitness evaluation is the robustness. In optimization, robustness may be discussed with respect to the object parameters or the condition parameters. The condition parameters are the parameters that define the working conditions under which the optimization is carried out. For example, a turbine blade is optimized at the Mach number of 1.0. Consider the minimization of the following function:

$$f(x) = g(x, a), \tag{2.18}$$

where a is a condition parameter, x is the object parameter, $g(\cdot)$ is a function of x. Assume function $g(x, a)$ has two minima with a fixed, as shown in Fig. 2.10(a). It can be seen that minimum A is more robust than minimum B with respect to object parameter x. However, the robustness of solution A and B is very different with regard to the condition parameter a, as illustrated in Fig. 2.10(b).

In both cases, more than one evaluation is needed to assign a fitness properly to an individual. Generally, averaging can be used. In other words, the function value on the current point $f(x)$ and a perturbed point $f(x + \Delta x)$

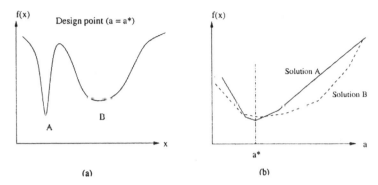

Fig. 2.10. Robustness of optimal solutions.

should be evaluated and the final fitness is the average of the function value of the two points:

$$f^i(t) = \frac{f(x) + f(x + \Delta x)}{2}, \qquad (2.19)$$

where i is the index for i-th individual, t is the generation index. The perturbation Δx can be set up deterministically or randomly. However, averaging needs re-sampling, which increases the number of fitness evaluation. A method that avoid extra fitness evaluations is to do averaging using generation based averaging. For example, in the t-th generation, the fitness of all individuals will be evaluated at point $x(t) + \Delta x(t)$ rather than at $x(t)$.

2.7 Parallel Implementations

Fitness evaluation is very time-consuming for some real world problems. One example is the computational fluid dynamics (CFD) simulation used in aerodynamic structure optimization, which takes from tens of minutes to several hours for one calculation. In this case, parallelization of the evolutionary algorithms becomes critical. The parallelization can either based on workstations connected by a network or on multi-CPU parallel computers. With the support of the related software, e.g., the Parallel Virtue Machine language [69], parallel implementation of evolutionary algorithms becomes feasible to most users.

The most popular scheme for parallel evolutionary algorithms is the global parallelization, see Fig. 2.11 which was implemented in [77]. This scheme is relatively simple and easy to carry out. Usually it adopts a synchronous master-slave structure, in which the master computer sends data to slave computers and receives data when the evaluation on the slave computers is finished. The evaluation can also be done asynchronously. In this case, a slave computer does not need to wait for another and can continue to do the next evaluation. In this case, there is no clear line between generations, which is also known as the steady-state GA or the steady-state ES [217, 233]. If there is shared memory available, the population can then be put in the memory and be accessed by each processor.

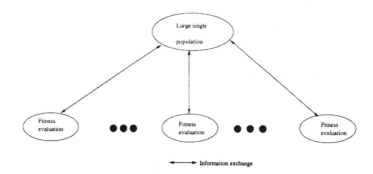

Fig. 2.11. The global model.

The second approach is the coarse grained parallelization, see Fig. 2.12. In this scheme, the whole population is divided into a number of sub-populations, which are also called demes [79]. Migration is introduced between different sub-populations for information exchange. In implementing migration, migration time-scale, migration rate and migration topology are three parameters to be determined [28]. The *island model* and the *stepping stone model* are two main models for migration topology. In the island model, migration can happen between any of the sub-populations, whereas in the stepping stone model, migration is restricted to the neighboring sub-populations. Coarse grained parallel evolutionary algorithms are sometimes called distributed evolutionary algorithms and can be extended to multi-objective evolutionary computation as one of the method for population-based non-Pareto approach.

If the whole population is divided into a large number of sub-populations, in extreme, each sub-population has only one individual. This is called fine grained parallelism and is suitable for large-scale parallel computers, refer to Fig. 2.13. Note that in the fine grained parallelization, communication becomes an important issue.

A hybridation of the above schemes is also possible. For example, the coarse grained model can be combined with the fine grained model or the global model.

2.8 Summary

In this chapter, the fundamentals of evolutionary computation are introduced. The main genetic operators, such as recombination, mutation and

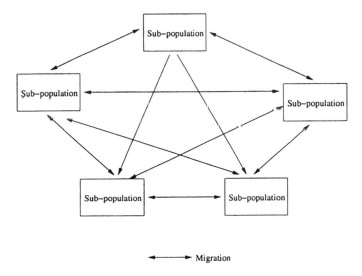

Fig. 2.12. The coarse grained model.

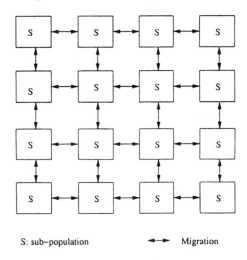

Fig. 2.13. The fine grained model.

selection are presented. Adaptation and self-adaptation used in genetic algorithms and evolution strategies are described. Important issues related to the fitness evaluation, such as constraints, multiobjectivity, uncertainties parallelization are discussed. These issues are of particular importance in dealing with real-world problems.

3. Artificial Neural Networks

3.1 Introduction

Artificial neural networks are historically simplified models for simulating the functions of human brains. From the cognition point of view, neural networks are able to acquire Knowledge through a process of learning and the acquired knowledge is stored in distribution among nodes and weights. Mathematically, most feed-forward neural networks are linear or non-linear function mapping systems that can perform function approximation (regression) and classification.

Neural networks have found a wide range of applications in computer science and almost all engineering sciences. In this applications, the function of the neural networks can generally be classified into three categories:

- Modeling of static or dynamic systems. Neural networks are able to realize a linear or nonlinear mapping. Therefore, it has widely been used for modeling, identification and time series prediction. To approximate a dynamic system, feedback neural networks can be used. Thanks to the learning ability of the neural networks, they have also been applied to adaptive control of nonlinear systems.
- Classification. Neural networks can memorize a number of patters in terms of weights. Classification can be seen as a discrete form of functional mappings.
- Optimization. Some feedback neural networks, for example, the Hopfield neural networks have been used in nonlinear dynamic programming to solve complicated optimization problems.

3.2 Feedforward Neural Network Models

Among the feed-forward neural networks, multi-layer perceptrons (MLP) and radial basis function networks (RBFN) are the two most commonly used feedforward artificial neural networks. Although the MLPs and the RBFNs are very similar in structure, there are also essential differences in several aspects.

3.2.1 Multilayer Perceptrons

Fig. 3.1 shows a multilayer perceptrons model. In this model, there are one
input layer with n input nodes, two hidden layers (with K and L hidden
nodes, respectively), and one output layer, with only one output node in this
example. The relationship between the input and output can be described by
the following equations:

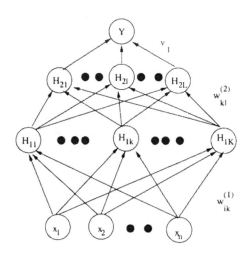

Fig. 3.1. Multilayer Perceptrons with n inputs, one output and two hidden layers.

$$H_{1k} = f(\sum_{i=1}^{n} w_{ik}^{(1)} x_i), \tag{3.1}$$

$$H_{2l} = f(\sum_{k=1}^{K} w_{kl}^{(2)} H_{1k}), \tag{3.2}$$

$$y = \sum_{l=1}^{L} v_l H_{2l}. \tag{3.3}$$

In the equations, $f(\cdot)$ is the activation function, which usually is the logistic
function or the tanh function. The nonlinear activation function is essential
to the power of MLPs, otherwise, it does not bring any benefit by combining
multiple nodes and layers. Besides, it is also useful to introduce a bias in the
hidden nodes, which can be realized by adding an input node with a constant
input value of 1. In this model, the output node is linear, which has proved
to be helpful for increasing the flexibility and improving the learning speed.
Strictly speaking, an MLP should have at least two hidden layers, however,

perceptron models that have one nonlinear hidden layer are used to be called MLP neural networks.

The mapping power of MLP networks has been mathematically demonstrated. It has been shown that MLPs with one hidden layer and an arbitrary number of hidden nodes are universal approximators [95]. The needed number of hidden nodes for an MLP to approximate a function with sufficient accuracy depends mainly on the properties of the function and the samples drawn from the function. In certain cases, an additional hidden layer can make it easier for a neural network to approximate a target function with fewer weight connections [33]. One disadvantage by using two hidden layers is that it may introduce extra local minima, which makes learning harder. Using evolutionary algorithms, it is possible to optimize number of layers and the number of hidden nodes of the neural network for a given task. For non-layered, more complicated structure optimization of neural networks, see [201].

3.2.2 Radial Basis Function Networks

Radial basis function networks (RBFN) are another often used neural network model [168, 182]. In contrast to MLPs, RBFNs usually have only one hidden layer. For an RBF network with n input nodes, H hidden nodes and one output node, the relationship between the input and the output can be described by:

$$y = \sum_{j=1}^{H} w_j \phi_j \left(\frac{\|\mathbf{x} - \boldsymbol{\mu}_j\|}{\sigma_j} \right), \tag{3.4}$$

where $\phi(\cdot)$ is the j-the radial basis function, $\boldsymbol{\mu}_j$ and σ_j are the center and the variance of the j-th basis function, and w_j is the weight connecting the j-th basis function and the output node. The most commonly used radial basis function is the Gaussian function, other functions such as the thin plate splines can also be used [76].

If the output of the RBF network can also be normalized, then we get

$$y = \frac{\sum_{j=1}^{H} w_j \phi_j(\|\mathbf{x} - \boldsymbol{\mu}_j\|/\sigma_j)}{\sum_{j=1}^{H} \phi_j(\|\mathbf{x} - \boldsymbol{\mu}_j\|/\sigma_j)}. \tag{3.5}$$

This normalized RBF network model has several interesting features [223]. The most interesting point is that the normalized RBF networks are mathematically equivalent to a class of fuzzy systems under certain minor conditions [108]. More discussions on this issue will be given in Chapter 7.

3.3 Learning Algorithms

Knowledge acquisition with the help of neural networks is realized by a process of *training* or *learning*. In most cases, knowledge of an unknown system is presented in the form of data. Therefore, most learning algorithms are data-driven. As it is well known, data collection is very difficult in some real-world problems. Thus, incorporation of a *prior* knowledge, including hints and expert knowledge [1, 123], into neural network becomes very important when available training data are insufficient.

Learning is realized by changing the parameters (weights and nodes) of the neural network to optimize an objective function. Depending on the involved learning method, the objective function can be very different. Generally, there are three main learning schemes.

3.3.1 Supervised Learning

In *supervised learning* , there is a teacher, i.e., a desired value for the neural network. Therefore, the objective of learning is to minimize the difference between the teacher signal and the neural network output. A large variety of supervised learning methods have been developed, mainly based on the gradient methods and their extensions, for example, the steepest descent method, the conjugate gradient method and the Levenberg-Marquardt method [16]. Learning algorithms based on *expectation maximization* have proved to be a class of very fast training algorithms [163]. Besides, evolutionary algorithms have played an important rule in neural network learning [248]. Detailed overviews of the recent advances on supervised learning can be found in [244, 111].

Very interestingly, the gradient-descent based learning method is also called *error back-propagation,* and the MLP networks using this learning method is called the *back-propagation neural network,* the BP network for short. This is ascribed to the fact that a similar method has been discover in [240], in which the error on the hidden layer is calculated by back-propagating the error on the output layer.

In the following, the learning algorithm based on gradient method is described. Suppose the desired output of the neural network in Fig. 3.1 is y_d, then the error function is defined by:

$$E = \frac{1}{2}(y_d - y)^2,$$ (3.6)

where y is the output of the network described by equations (3.1-3.3). According to the gradient method, the weights of the network can be updated in the following way by minimizing the error function:

$$v_l(t+1) = v_l(t) - \xi \frac{\partial E}{\partial v},$$ (3.7)

$$w_{kl}^{(2)}(t+1) = w_{kl}^{(2)}(t) - \xi \frac{\partial E}{\partial w_{kl}^2},$$ (3.8)

$$w_{ik}^{(1)}(t+1) = w_{ik}^{(1)}(t) - \xi \frac{\partial E}{\partial w_{ik}^1},$$ (3.9)

where $i = 1, ..., n$, $k = 1, ..., K$ and $l = 1, ..., L$, ξ is a small positive constant called *learning rate*. By combining equation (3.3) with equation (3.7), it is straightforward that:

$$\frac{\partial E}{\partial v_l} = \frac{\partial E}{\partial y} \frac{\partial y}{\partial v_l}$$

$$= (y_d - y) H_{2l}. \tag{3.10}$$

Thus, we can easily get the learning algorithm for the weights connecting the output layer and the second hidden layer:

$$v_l(t+1) = v_l(t) - \xi (y_d - y) f(\sum_{k=1}^{K} w_{kl}^{(2)} f(\sum_{i=1}^{n} w_{ik}^{(1)} x_i)). \tag{3.11}$$

To derive the learning algorithm for $w_{kl}^{(2)}$, the chain rule can be applied:

$$w_{kl}^{(2)}(t+1) = w_{kl}^{(2)}(t) - \xi \frac{\partial E}{\partial y} \frac{\partial y}{\partial H_{2l}} \frac{\partial H_{2l}}{\partial w_{kl}^{(2)}}$$

$$= w_{kl}^{(2)}(t) - \xi (y_d - y) v_l f'(\cdot) H_{1k}, \tag{3.12}$$

where $f'(\cdot)$ is the derivative of the activation function. Define

$$\delta_l^{(2)} = (y_d - y) v_l f'(\cdot), \tag{3.13}$$

then we get

$$w_{kl}^{(2)}(t+1) = w_{kl}^{(2)}(t) - \xi \delta_l^{(2)} H_{1k}. \tag{3.14}$$

Similarly, we can apply the chain rule on $w_{ik}^{(1)}$:

$$w_{ik}^{(1)}(t+1) = w_{ik}^{(1)}(t) - \xi \frac{\partial E}{\partial y} \sum_{l=1}^{L} \frac{\partial H_{2l}}{\partial H_{1k}} \frac{\partial H_{1k}}{\partial w_{ik}^{(1)}}$$

$$= w_{ik}^{(1)}(t) - \xi (y_d - y) \sum_{l=1}^{L} v_l f'(\cdot) w_{kl}^{(2)} f'(\cdot) x_i. \tag{3.15}$$

By defining

$$\delta_k^{(1)} = \sum_{l=1}^{2} w_{kl}^2 f'(\cdot) \delta_l^{(2)}, \tag{3.16}$$

we have the following learning algorithm for $w_{ik}^{(1)}$:

$$w_{ik}^{(1)}(t+1) = w_{ik}^{(1)}(t) - \xi \delta_k^{(1)} x_i. \tag{3.17}$$

From equations (3.11), (3.3.1) and (3.17), it can be seen that supervised learning can be written in the following generic form:

$$\Delta w = -\xi \delta x. \qquad (3.18)$$

In the equation, x is the input of the neuron, δ is the error on the output side of the neural. For an output neuron, the error can be directly calculated by comparing the desired output and the output of the neuron. For a neuron in a hidden layer, the error can be calculated by back-propagating errors at the output layer.

When there is more than one training data , both the *batch-mode* and the *pattern-mode* learning can be used. In the batch-mode learning, the derivative for each data sample is summed up and then applied for weights adaptation. The pattern-mode learning is also known as *on-line* learning, that is, weights adaptation is carried out for each data sample iteratively.

The above standard learning algorithm may converge very slowly. There are a lot of methods for improving the standard gradient-based learning method, including the adaptation of the learning rate and the introduction of a momentum term. For details, refer to [189].

3.3.2 Unsupervised Learning

In the *unsupervised learning*, there is usually no teacher signal. However, in most unsupervised learning algorithms, the input signal is used as the teacher. The typical tasks for the unsupervised learning are self-organization [140], association and classification [7]. The unsupervised learning is very useful for clustering the location (center) of the basis functions in the RBF networks, which has shown to be very efficient in some cases for the training of the RBF networks [245].

Hebbian learning and *competitive learning* are two basic   unsupervised learning algorithms. In the Hebbian learning, the weight increases if the input and output of a neuron changes synchronously and decreases if asynchronously:

$$\Delta w = \xi y x, \qquad (3.19)$$

where x and y are the input and output of the neuron.

In competitive learning, there are a number of neurons in the output layer that are competing among themselves. The neuron with the maximal output is said to win and is activated, and all other neurons are deactivated. Therefore, the competitive learning is also called the *Winner-Take-All*. Suppose the j-th output neuron is the winner, see Fig. 3.2, then the weights connecting to this winning neuron can be updated by

$$\Delta w_{ij} = \xi (x_i - w_{ij}), \qquad (3.20)$$

where i means the i-th input. All other weights remain unchanged.

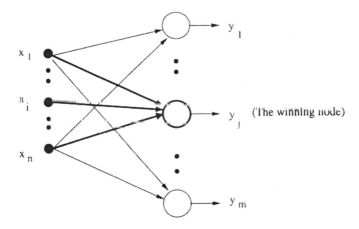

Fig. 3.2. The competitive learning.

3.3.3 Reinforcement Learning

In the *reinforcement learning*, the output of a neural network is treated as an action based on some instructions. This action will change the state of the environment. If the change of the state is desired, a *positive reward* will be assigned to the action; if the change of the state is not desired, then a *negative reward* will be assigned to the action. In this way, the neural network is able to learn to behave better, although the desired action is unknown. *Temporal Difference Learning* [214] and *Q-learning* [237] may be the two most influential reinforcement learning algorithms. A good overview over this field can be found in [132].

Temporal Difference Learning. Assume the current state is s, the current action is $a = \pi(s)$ and the next state is s'. Then the value function of the action π is

$$f^\pi(s) = (1 - \alpha) \cdot f^\pi(s) + \alpha \cdot [R(s, a, s') + \gamma f^\pi(s')], \qquad (3.21)$$

where, α is the learning rate, typically between 0.01 and 0.5, $R(s, a, s')$ is the immediate reward function. This algorithm is known as $TD(0)$, which no explicit model is used for computing the policy value.

In order to avoid storing a separate value for each state s, the value function can be estimated using an approximate model, e.g., a neural network. Suppose the approximate model is in the form of $f^\pi(s, W)$, where W is a vector of weights that can be adjusted in such a way that the model is able to approximate the true value $f^\pi(s)$. Thus, the following error function can be defined:

$$J = \frac{1}{2} \left(f^\pi(s, W) - [R(s, a, s') + \gamma f^\pi(s', W)] \right)^2, \qquad (3.22)$$

which is called the squared difference between $f^\pi(s, W)$ and the estimated cost based on the simulated outcome of the current stage. Thus, a learning rule for W can be obtained:

$$\Delta W = -\alpha \nabla_W f(s, W)\, (f^\pi(s, W) - [R(s, a, s') + \gamma f^\pi(s', W)])\,, \quad (3.23)$$

where α is the step-size, $\nabla_W f(s, W)$ is the gradient of $f(s, W)$ with respect to W.

Suppose the system have visited a sequence of states $s_1, s_2, \ldots, s_t, s_{t+1}$, then the afore-mentioned learning algorithm can be generalized by introducing an eligibility trace:

$$\Delta W_t = -\alpha z_t\, (f^\pi(s_t, W_t) - [R(s_t, a, s_{t+1}) + \gamma f^\pi(s_{t+1}, W_t)])\,, \quad (3.24)$$
$$z_{t+1} = \lambda z_t + \nabla_{W_t} f(s_t, W_t), \quad (3.25)$$

where z_t is called the eligibility coefficient.

Q-learning. If no explicit model of the system is available, the Q-learning method can be used. Let us define an action-value function $Q(s, a)$ to replace the value function $f(s)$, where $Q(s, a)$ is the expected cumulative discounted reward of performing action a in a state s and then the current action. Thus, the value of a state is the maximum of the Q values for that state:

$$f(s) = \max_a Q(s, a). \quad (3.26)$$

The Q version of the Bellman's equation can be defined as follows:

$$Q(s, a) = \sum_{s'} p(s'|s, a)[R(s'|s, a) + \gamma \max_{a'} Q(s', a')], \quad (3.27)$$

where $p(s'|s, a)$ is the probability of state s' for a given state s and an action a. In term of the Q-factors, the value iteration algorithm can be written as:

$$Q(s, a) = \sum_{s'} p(s'|s, a)[R(s, a, s') + \gamma \max_a Q(s', a)]. \quad (3.28)$$

A more general version of this iteration is:

$$Q(s, a) = (1 - \gamma)Q(s, a) + \alpha\, (R(s, a, s') + \gamma \max_a Q(s', a))\,. \quad (3.29)$$

Proofs of the convergence of the temporal difference algorithm and the Q-learning algorithm are discussed in detail in [16].

3.4 Improvement of Generalization

Neural networks can learn training samples perfectly, but not necessarily have good performance on the samples that have not been trained. The ability for the neural networks to give correct outputs on the inputs that are not in the training set is call the *generalization* of the neural network. Naturally, the generalization ability is of great concern for neural networks users.

One point that needs to be pointed out is that the generalization ability of the neural networks not only depends on the neural network model and the learning method used, but also on the information contained in the training samples about the function to learn and the target function itself. One should make sure that the samples used to train the neural network contains sufficient information about the target function. This requires a sufficient number of training data, which is not always practical. Thus, use of heuristics or expert knowledge may be critical for a good generalization performance of neural networks.

A rigorous discussion about the generalization of neural network was first carried out in [70]. It is pointed out that the generalization error is affected by two factors: *bias and variance*. To get a good generalization, the neural network needs to make a trade-off between the bias and the variance. A large number of methods have been developed to improve the generalization of neural networks, which can be largely divided into five classes.

3.4.1 Heuristic Methods

Early stopping and training with noise (*jitter*) are two heuristic methods attempting to improve the generalization ability of neural networks. In the early stopping method, available data are divided into a *training set* and a *validation set*. Instead of training the network to reduce the error on the training set as much as possible, the training is stopped when the error on the validation set begins to increase. Issues such as how to split the data and when to stop need to be addressed carefully [234, 183].

By jitter, it is meant to deliberately add random noise to the inputs of training samples. It is believed that random noises can smooth the target function (the function itself is smooth, it may look rugged because of the sparse sampling) and thus prevent the network from over-fitting the training data. A closer look at this issue can be found in [93].

3.4.2 Active Data Selection

The purpose of *active data selection* is twofold. When training data are given, active data selection during learning is able to improve the learning performance significantly. Usually, active data selection can be done by optimizing an objective function, e.g., to maximize information gain. Another approach is to re-sample the training data, among which the *leave-one-out* is one of the most commonly used methods.

What seems to be more interesting is to select data actively in generating training data so that the learning performance can be enhanced with only a small data set. These methods are often used in the field of *design of experiments* (DOE) [171] to reduce the number of needed data when data collection is expensive. *Orthogonal Array* and *D-optimality* are two widely used methods. Let us first consider the orthogonal array method. Assume N experimental runs are conducted for a system that has k inputs and one output. If a first-order polynomial model is used to approximate the system, the orthogonal array method states that the variance of estimate is minimized if

$$
\begin{bmatrix}
1 & x_{11} & x_{12} & \cdots & x_{1k} \\
1 & x_{21} & x_{22} & \cdots & x_{2k} \\
\vdots & \vdots & \vdots & & \vdots \\
1 & x_{N1} & x_{N2} & \cdots & x_{Nk}
\end{bmatrix}
\tag{3.30}
$$

is orthogonal. A special class of first-order orthogonal design is called *simplex designs*, in which $N = k + 1$ data sets

$$
\mathbf{X} =
\begin{bmatrix}
x_{11} & x_{12} & \cdots & x_{1k} \\
x_{21} & x_{22} & \cdots & x_{2k} \\
\vdots & \vdots & & \vdots \\
x_{N1} & x_{N2} & \cdots & x_{Nk}
\end{bmatrix}
\tag{3.31}
$$

are collected. Then the simplex design requires that the angle that any two points make with the origin is θ, where

$$
\cos(\theta) = -\frac{1}{k}.
\tag{3.32}
$$

D-optimality is the best known criterion in experimental designs. It states that the experimental points should be chosen to maximize the determinant of the *moment matrix*

$$
\mathbf{M} = \frac{\mathbf{X}^T \mathbf{X}}{N}.
\tag{3.33}
$$

Refer to [171] for more discussions on theory and practice of design of experiments.

3.4.3 Regularization

Structural Regularization. The basic idea of *structural regularization* is that the neural network will not over-fit the training data if the complexity of the neural network is comparable to that of the target function. *Pruning* methods [190] and growing methods [54] have been widely studied to control the complexity of neural networks. In the pruning methods, a large initial network is trained and then some of weights or nodes are deleted. Smallest weight, smallest connection variance, smallest connection/node sensitivity, *Optimal Brain Damage* [154], *Optimal Brain Surgeon* [86] and minimal description length are the most widely used criteria for pruning.

In the growing methods, a small neural network is first used. Additional layers and nodes are added if necessary until a satisfying solution is found. Growing methods are also called constructive methods.

Formal Regularization. In formal regularization, an extra penalty term is added to the cost function of the training algorithm

$$J = E + \lambda \Omega, \tag{3.34}$$

where, E is the conventional error function, $0 \le \lambda < 1$ is the regularization coefficient, Ω is the regularization term. Mainly, three types of approaches to formal regularization are available. The first approaches are the heuristic approaches, in which the penalty term smoothes the curvature, or penalizes large weights. The most popular heuristic regularization method is called *weight decay*. For example, we can use the following regularization term [147]

$$\Omega = \frac{1}{2} \sum_i w_i^2. \tag{3.35}$$

One problem of the above exponential weight decay is that the neural network may use many small weights. Another weight decay method is proposed as follows [239]:

$$\Omega = \sum_i \frac{(w_i/w_0)^2}{1 + (w_i/w_0)^2}, \tag{3.36}$$

where w_0 is a pre-defined value. A combination of these two methods has also been suggested in [204]

$$\Omega = \frac{1}{2} \sum_i w_i^2 + \eta \sum_i \frac{(w_i/w_0)^2}{1 + (w_i/w_0)^2}. \tag{3.37}$$

An alternative to the exponential decay is

$$\Omega = \sum_i |w_i|. \tag{3.38}$$

This method has an advantage that the decay is constant and is able to drive insignificant weights to zero, which is very useful in rule extraction from neural networks.

The second approach, based on information theory, uses *mutual information* to resist the excessive decorrelation of the hidden nodes so that overfitting can be alleviated [47]. Suppose a neural network has N hidden nodes $y_j, j = 1, .., N$. Before the mutual information can be computed, y_j needs to be normalized so that it can be treated as the conditional probability $p(C_j|x^k)$, i.e., the probability of a given input pattern x^k belongs to class C_j. Then the regularization term using the mutual information between inputs and the hidden nodes:

$$\Omega = \sum_j MI_{x,y_j}. \tag{3.39}$$

In the equation, MI_{x,y_j} is the definition of mutual information between the input and the j-th hidden node:

$$MI_{x,y_j} = -\bar{y}_j \log[\bar{y}_j] + \frac{1}{N} \sum_{n=1}^{N} y_j^n \log[y_j^n], \tag{3.40}$$

where $\bar{y}_j$ is the average over the input patterns, N is the number of input patterns and y_j^n is the j-th hidden node given the n-th input pattern.

The third approaches are the Bayesian approaches, in which the penalty term can play a role of the *a priori* distribution of the weights. Gaussian regularizer, Laplacian regularizer, Cauchy regularizer and a mixture of Gaussians can be employed. A detailed description of the formal regularization is provided in [18].

All the methods for improving the generalization property of neural networks are useful in many cases. Nevertheless, some researchers believe that over-fitting is not as serious as it is expected. Moreover, it is argued that small-size networks are hard to train and large networks may help avoid local minima and improve generalization [224, 152].

It should be pointed out that the regularization coefficient ξ plays an important rule in learning with regularization. If ξ is too large, the learning capability of the neural network may be seriously harmed and the learning becomes difficult.

3.4.4 Network Ensembles

The concern between the bias and variance can also be balanced by constructing network *ensembles* (also called *committee*). Instead of building one single network model, a number of networks can be constructed and then be combined into one output. The ensembles can be obtained by manipulating the training data based on statistic methods. *Bagging* (bootstrap aggregation) [25] and *Boosting* [67] (improved version of boosting developed in [196]) are two popular methods to improve the performance of the neural networks.

The basic idea of bagging is to generate an ensemble of estimators, each using a bootstrap sample of the original training data set. Assume the original data set is $X = (x_1, x_2, ..., x_n)$, bootstrap sampling means to create a new data set by uniformly sampling n instances from X with replacement. For each bootstrap sample $(X^*(t), i = 1, ..., B)$, an estimators NN^i can be constructed. The final output of the model is obtained by averaging $B + 1$ estimators

$$NN = \frac{(NN^0 + NN^1 + ... + NN^B)}{B + 1}, \tag{3.41}$$

where NN^0 is the estimator built on the original data set X.

While bagging averages each estimator without weighting, boosting maintains a weight for each bootstrap data set - the higher the weight, the stronger influence the sample has. The final output is a weighted aggregation of the estimators, with each weight being a function of the accuracy of the corresponding estimator. Let $D_i(0) = 1/n, i = 1, ..., n$ the initial distribution for bootstrap sampling. At iteration t, a data set $X^*(t)$ is generated according to distribution $D_i(t)$ and then an estimator $NN(t)$ is built. The error $\epsilon(t)$ of this estimator can be obtained. Then the weight is calculated by.

$$w(t) = \frac{1}{2}\ln\left(\frac{1 - \epsilon(t)}{\epsilon(t)}\right),$$ (3.42)

and the distribution is adjusted by

$$D_i(t+1) = D_i(t) \times \begin{cases} e^{-w(t)} & \text{if pattern } i \text{ is correct classified,} \\ e^{w(t)} & \text{if pattern } i \text{ is incorrect classified.} \end{cases}$$ (3.43)

The distribution $D_i(t+1)$ needs to be normalized so that $\sum_{i=1}^{n} D_i(t+1) = 1$. The final output of the estimator is a weighted aggregation of B estimators

$$NN = \sum_{t=1}^{B} w(t)NN(t).$$ (3.44)

Note that $\epsilon(t)$ is assumed to be less than 0.5. In classification, it means that classifier $NN(t)$ is better than a random classifier. It has been proved that boosting is able to reduce both bias and variance and therefore will improve the generalization capability.

An ensemble of network models can also be constructed by adding a term to the error function that penalizes the correlation between the ensembles. Refer to [160] for more details.

3.4.5 *A Priori* Knowledge

All the aforementioned methods are based on the implicit assumption that the target function is smooth, which is not necessarily true. On the other hand, it will be very helpful if there is a priori knowledge about the target function. A priori knowledge can include hints and expert knowledge [1, 123]. Besides, knowledge in terms of automaton rules has also been considered [64] to be integrated into recurrent networks. The neural network consists of two subnetworks, one with randomly initialized weights, and the other with the weighted initialized based on the automaton rules using a linear programming technique.

3.5 Rule Extraction from Neural Networks

Although a neural network can acquire knowledge from data, it is difficult for human beings to understand the knowledge distributed among the weights and nodes of the neural networks. An effort to address this problem is to extract symbolic or fuzzy rules from the trained neural networks. The main benefit from extracting rules from neural networks is that one can gain deeper insights into the networks to verify the internal logic of the system on the one hand, and to discover new dependencies in the training data on the other hand.

Main issues that need to be taken into account in extracting rules from neural networks are the fidelity and the comprehensibility of the extracted knowledge. Fidelity means that the extracted rules are able to capture all of the information embedded in the original neural network and do not add extra information that the neural network does not contain. Comprehensibility concerns about the interpretability of the extracted rules. If the extracted rules are not understandable to human beings, one can not benefit much from rules extraction. Interpretability of the extracted rules will be further discussed in the context of extracting fuzzy rules from data and RBF networks in Chapter 7.

Techniques to extract rules from neural networks can generally be divided into *de-compositional approaches*, *pedagogical approaches* and a combination of both approaches [2, 226]. In the de-compositional approach, a node is translated into a rule, in which the inputs of the node are the condition part and the outputs are the consequence part of an extracted rule. Obviously, the extracted rules using this method may have good fidelity, however, the comprehensibility of the extracted rules may be harmed, for example, when extracting rules from networks with more than one hidden layer.

The pedagogical approach treats the neural network as a black box and view rule extraction as a learning task. The structure of the rules may be prescribed and their parameters are learned using data samples generated from the neural network.

3.5.1 Extraction of Symbolic Rules

The basic idea in the earlier work on symbolic rule extraction is to find sets of weights that can activate the neuron regardless the values of other connections. Consider the j-th hidden neuron in Fig. 3.3 with a threshold activation function and Boolean inputs (0 or 1), it is easy to extract the following symbolic rules for the j-th neuron:

$$\text{If } x_1 \text{ and } x_2 \text{ and } x_3, \text{ then } y_j;$$
$$\text{If } x_1 \text{ and } x_2 \text{ and } \neg x_4, \text{ then } y_j;$$
$$\text{If } x_1 \text{ and } x_3 \text{ and } \neg x_4, \text{ then } y_j.$$

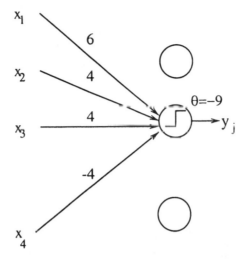

Fig. 3.3. A neural network for rule extraction.

Several rule extraction methods have been developed from this basic rule extraction scheme, e.g., the *M-of-N* technique that extracts rules in the following form:

If (*M* of the following *N* antecedents are true), then y_j.

More symbolic rule extraction methods can be found in [2].

In order to reduce the number of extracted rules and to improve the understandability of the rules, it is necessary to use regularization techniques such as pruning and weight decay. For example, a three-phase rule extraction algorithm is proposed in [204]. At first, neural network training with weight decay based on equation (3.36) is carried out. Then, the insignificant weights are removed. Finally, rules are extracted by discretizing the hidden node activation values through clustering. By analyzing the relationship between the clusters center and the output, rules in the form of

$$\text{If } x \leq x_1^0 \text{ and } ... \text{ and } x_n \leq x_n^0, \text{ then } y \text{ is } C,$$

where $x_i, i = 1, ..., n$ are the inputs, C is one class of the output. Similar work using the regularization in equation (3.38) and pruning is also reported in [105].

3.5.2 Extraction of Fuzzy Rules

Extraction of fuzzy rules or more strictly, fuzzy-like rules has also widely been investigated. An interesting approach to the interpretation of MLP neural networks is proposed in [32]. The basic idea is to decompose the sigmoid activation function into three TSK fuzzy rules, see Fig. 3.4. Thus, a TSK

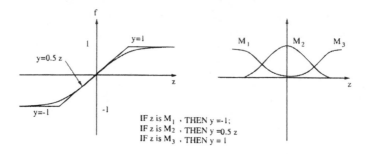

IF z is M_1 , THEN y =-1;
IF z is M_2 , THEN y =0.5 z
IF z is M_3 , THEN y = 1

Fig. 3.4. Decomposition of a sigmoid function.

fuzzy rule can be generated for each hidden neuron. For a given input pattern, a rule similar to the TSK fuzzy rule can be extracted from each output neuron by combining the TSK fuzzy rules of the hidden neurons connecting to it. Consider a neural network with n inputs, two hidden neurons and one output neuron, refer to Fig. 3.5. For a given input pattern, hidden neuron 1 generates the rule "IF z is M_1, THEN $y_1 = -1$", and hidden neuron 2 generates the rule "IF z is M_2, THEN $y_2 = 0.5z$. Therefore, the following fuzzy-like rule can be extracted from the output neuron:

$$\text{IF } [x_1, ..., x_n] \in M_1 \cdot M_2, \text{ THEN } y = w_1 y_1 + w_2 y_2 + \theta.$$

Although such a fuzzy rule provides an explanation of the neural network, it is difficult to interpret it.

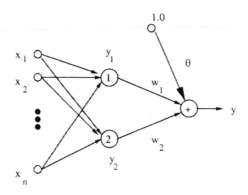

Fig. 3.5. A neural network for fuzzy rule extraction.

Another attempt to interpret an MLP network with fuzzy rules was reported in [12]. By developing a new fuzzy operator called *interactive-OR*, the authors succeeded in converting the MLP neural network directly to a fuzzy rule system. The most significant feature of this method is that the extracted fuzzy system and the neural network are completely equal and the parameters of the fuzzy rule are directly related to those of the neural network. The fuzzy rules look like

IF $x_1 w_{1j}$ is A_j $\star x_2 w_{2j}$ is A_j $\star$ ... $x_n w_{nj}$ is A_j, THEN y is p_j,

where j means the j-th hidden neuron, p_j is the weight connecting the j-th hidden neuron and the output neuron, and $\star$ is the interactive-OR operator

$$\star(a_1, a_2, a_n) = \frac{a_1 \cdot a_2 \cdot ... \cdot a_n}{(1 - a_1) \cdot (1 - a_2) \cdot ... \cdot (1 - a_n) + a_1 \cdot a_2 \cdot ... \cdot a_n}. \quad (3.45)$$

It is easy to verify that the premise of the fuzzy rule combined by the interactive-OR operator is equivalent to $\sum_{i=1}^{n} x_i w_{ij}$ is A. Thus, If fuzzy set A_j is defined by the sigmoid function of the j-th hidden neuron, the fuzzy rule is equivalent to a hidden neuron with the following input-output relation

$$y_j = p_j f_j \left(\sum_{i=1}^{n} w_{ij} x_i \right), \quad (3.46)$$

where f_j is the sigmoid function of the j-th hidden neuron.

In spite of the great success they have achieved, difficulties arise in understanding the fuzzy rules. First of all, *interactive-or* is neither a fuzzy AND nor a fuzzy OR operator. Secondly, the authors assumed that "$x_i w_{ij}$ is A_j" might be interpreted as "x_i is A_j/w_{ij}" and "$x_i w_{ij} + \theta_j$" might be interpreted as "x_i is $(A_j - \theta_j)/w_{ij}$", where x_i is the input, w_{ij} is the input weight, θ_j is the bias of the neuron divided by the number of the inputs and A_j is a fuzzy set described by the sigmoid function of the neuron. Unfortunately, this assumption is not straightforward because of the nonlinearity of the membership function.

As a conclusion, it is difficult to extract interpretable fuzzy rules from standard MLP networks using decompositional approaches. One method to avoid this difficulty is to use neural network structures that can directly converted to fuzzy rules, which are usually called neurofuzzy systems [109]. In addition, it seems more natural to extract fuzzy rules from RBF networks [127], which will be presented in Chapter 7.

3.6 Interaction between Evolution and Learning

Evolution and learning are the two most important adaptation mechanisms in the nature, from which evolutionary and learning algorithms have been developed. Naturally, the interaction between evolution and learning has also received increasing attention in the community of evolutionary computation.

Two main phenomena embodying the interaction between evolution and learning found in nature are the *Baldwin effect* and the *Lamarckian theory*. In the Lamarckian theory, it is believed that the characteristics learned (acquired) by individuals during their lifetime can be directly inherited. While the Darwinism advocated evolution in small and incremental steps, the Lamarckism expected occasional rapid changes. However, the Lamarckian theory was called in question when it was shown that the information acquired is unlikely to be able to directly transmitted to the genetic coding. In the field of artificial evolutionary and learning systems, the Lamarckian mechanism has found successful applications in the optimization of neural networks [24, 249], in which the weights of neural networks adjusted by local learning are encoded back in the chromosome. It was argued that the introduction of the Lamarckian mechanism may cause instability [194], which can be alleviated in a way that the acquired new information is "averaged" before it is transmitted to the genetic coding [98].

On the other hand, the Baldwin effect suggested that the learned characteristics can be indirectly inherited. Individuals that are able to learn characteristics during their lifetime that are beneficial to survival in evolution will have larger chance to be selected and passed to the next generation. When the population is overwhelmed by individuals that are able to learn the characteristics, they may be assimilated into genetic coding in a large time scale. The Baldwin effect has been studied in the artificial evolutionary and learning systems [90, 10].

Besides, it has also been suggested that evolution on one task may benefit from the lifetime learning on another unrelated or weakly related task [176]. This effect was further investigated in [85] and was called another "new factor" in evolution.

Integration of learning or other local search strategies in evolution is sometimes known as the *memetic algorithms* [169]. Interesting discussions on the Lamarckian mechanism and the Baldwin effect is also provided in [242].

Nevertheless, the interaction between evolution and learning in natural and artificial systems is far from fully investigated. *A priori* knowledge or expert knowledge, which can be considered as a kind of indirect learning, may also play an important rule in evolution [126]. In most studies, the attention is focused on the benefits that evolution can gain from learning. In fact, it is also found [121] that information accumulated in evolution can also be used to guide learning. Systematic comparative studies on the differences between the Lamarckian mechanism and the Baldwin effect in artificial evolutionary and learning systems are needed.

3.7 Summary

The fundamentals of artificial neural network theory are introduced. Two widely used feedforward neural networks models, the MLP and the RBFN are described. Three main learning mechanisms, supervised learning, unsupervised and reinforcement learning are briefly presented. Methods for improving the generalization capability of feedforward neural networks for regression and classification are discussed. To improve the interpretability of neural networks, methods for rules extraction from trained neural networks are illustrated. Finally, the interaction between evolution and learning is discussed.

4. Conventional Data-driven Fuzzy Systems Design

4.1 Introduction

The central point in designing a fuzzy system for modeling and control is the generation of a set of efficient and effective fuzzy rules. Generally, fuzzy rules can be generated either from heuristics or from experimental data. Heuristic fuzzy rule generation, as discussed in Chapter 1, consists of the following main steps:

- Determination of the structure of the fuzzy systems. This includes mainly the selection of the inputs and outputs. Usually, a multi-input multi-output fuzzy systems can be decomposed into a number of multi-input single-output fuzzy systems.
- Definition of linguistic terms and their fuzzy membership function for each fuzzy variables. If a Mamdani fuzzy system is designed, each input and output can be seen as a linguistic variable. Therefore, it is necessary to define the linguistic terms (e.g., Small, Medium, Large etc.) and their fuzzy membership function. Details about the definition of a linguistic variable is presented in Section 1.2.1.
- Determination of fuzzy inference method. For example, for fuzzy mapping rules, the fuzzy Cartesian product can be used to define the fuzzy relation. Refer to Section 1.2.2.
- Defuzzification. If the Mamdani type fuzzy rules are used, the fuzzy output derived from the fuzzy rule base should be defuzzified. Refer also to Section 1.2.2 for different defuzzification methods. If the Takagi-Sugeno fuzzy rules are used, the defuzzification is included in the fuzzy inference and a crisp output is obtained directly.

The main weakness of the heuristic fuzzy rule generation method lies on the one hand in its difficulty in generate fuzzy rules for high-dimensional systems. It is well known that human thinking is usually no more than three-dimensional. Thus, if the system to be modeled or controlled is high-dimensional, it is very difficult for human experts to abstract sophisticated fuzzy rules from experience.

Since the beginning of 1990's, data-driven fuzzy rule generation has been widely investigated and several methods have been developed. The main advantage of data-driven fuzzy rule generation methods is that they are able to generate fuzzy rules for unknown systems without any a priori knowledge, provided that a number of data representing the input-output relation of the system are available. In this way, it is expected that knowledge about the unknown system can be acquired from the fuzzy rules. One important condition is that the fuzzy rules generated from data are interpretable, that is, transparent to human users, as it has been discussed in Section 1.3.

In the following sections, three conventional methods for generating fuzzy rules from data will be presented. The first method takes advantage of the fuzzy inference based on the fuzzy relation operations. In the second method, fuzzy rules are generated from data in five steps [236]. The third method generates a fuzzy rule for each data pair, assuming the Gaussian functions are used for fuzzy membership functions.

Unlike heuristic fuzzy rule generation, the following issues must be addressed in the data-driven fuzzy rule generation, which are related to the main aspects of interpretability of fuzzy rules.

- The number of fuzzy subsets in a fuzzy partition should not be too large. A very large number of fuzzy subsets makes it hard to assign meaningful linguistic terms to the fuzzy subsets.
- The number of fuzzy rules in the rule base should be too large. It happens easily if the number of available data pair is large.
- The consistency of the fuzzy rules. Unlike the heuristic approach to fuzzy rules generation, fuzzy rules generated from data may conflict to each other seriously. The conflicts can be resolved either by assigning a truth degree to each fuzzy rule [236] or by checking the consistency of the fuzzy rules [126].

4.2 Fuzzy Inference Based Method

Suppose the following input-output data pairs are given for a two-input single-output system:

$$D^{(i)} = (x_1^{(i)}, x_2^{(i)}, y^{(i)}), i = 1, 2, \cdots, N, \tag{4.1}$$

where x_1 and x_2 are inputs, y is the output, and N is the number of available data pairs. The method for generating fuzzy rules based on fuzzy inference consists of seven steps:

1. Define the fuzzy linguistic terms and their membership functions. Suppose x_1 has 6 linguistic terms $\{A_1^1, A_1^2, A_1^3, A_1^4, A_1^5, A_1^6\}$, x_2 has 5 linguistic terms $\{A_2^1, A_2^2, A_2^3, A_2^4, A_2^5\}$ and y has 3 linguistic terms $\{B^1, B^2, B^3\}$.
2. For each data pair D^i, calculate the fuzzy set of each variable:

$$X_{1i} = \frac{\mu_{A_1^1}(x_{1i})}{A_1^1} + \frac{\mu_{A_1^2}(x_{1i})}{A_1^2} + \cdots + \frac{\mu_{A_1^7}(x_{1i})}{A_1^7}, \tag{4.2}$$

$$X_{2i} = \frac{\mu_{A_2^1}(x_{2i})}{A_2^1} + \frac{\mu_{A_2^2}(x_{2i})}{A_2^2} + \cdots + \frac{\mu_{A_2^5}(x_{2i})}{A_2^5}, \tag{4.3}$$

$$Y_i - \frac{\mu_{B^1}(y_i)}{B^1} + \frac{\mu_{B^2}(y_i)}{B^2} + \frac{\mu_{B^3}(y_i)}{B^3}, \tag{4.4}$$

where $\mu(\cdot)$ is the membership function of the corresponding fuzzy set, X_1, X_2 and Y are fuzzy sets on the universe of discourse of x_1, x_2 and y.

3. Calculate the fuzzy relation between the inputs:

$$R_{X_{2i} \times X_{1i}} = \begin{bmatrix} \mu_{A_2^1}(x_{2i}) \\ \mu_{A_2^2}(x_{2i}) \\ \vdots \\ \mu_{A_2^5}(x_{2i}) \end{bmatrix} \wedge [\mu_{A_1^1}(x_{1i}), \mu_{A_1^2}(x_{1i}), \cdots, \mu_{A_1^7}(x_{1i})]. \tag{4.5}$$

4. Calculate the fuzzy relations between the inputs and output:

$$R_{y,B^1,i} = \mu_{B^1}(y_i) \wedge R_{X_{2i} \times X_{1i}}, \tag{4.6}$$

$$R_{y,B^2,i} = \mu_{B^2}(y_i) \wedge R_{X_{2i} \times X_{1i}}, \tag{4.7}$$

$$R_{y,B^3,i} = \mu_{B^3}(y_i) \wedge R_{X_{2i} \times X_{1i}}. \tag{4.8}$$

5. Combine the fuzzy relation between the inputs and output of the N data pairs:

$$R_{y,B^1} = \bigcup_{i=1}^{N} R_{y,B^1,i}, \tag{4.9}$$

$$R_{y,B^2} = \bigcup_{i=1}^{N} R_{y,B^2,i}, \tag{4.10}$$

$$R_{y,B^3} = \bigcup_{i=1}^{N} R_{y,B^3,i}. \tag{4.11}$$

6. Select an appropriate threshold $\lambda \in [0,1]$ and perform the λ-cut operation. This means that in the fuzzy relation matrix, if an element is larger than λ, then this element is set to 1; otherwise, it is set to 0. After that, each element with the value of 1 is replaced with the corresponding fuzzy subset. For example, all elements with the value of 1 in R_{y,B^1} are replaced with fuzzy subset B^1.

7. Add up all the fuzzy relations to get the final fuzzy relation, which contains all fuzzy rules generated from the given data.

To illustrate this method, let us consider the following example. Two data pairs are given as follows:

$$D^{(1)} : (0.75, 0.25, 0.2),$$
$$D^{(2)} : (0.9, 0.8, 0.75).$$

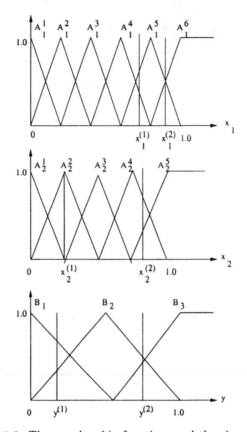

Fig. 4.1. The membership functions and the given data.

The membership functions for x_1, x_2 and y are shown in Fig. 4.1. Therefore, we can transfer the give data pairs to the following two fuzzy sets:

$$X_{11} = (0, 0, 0, 0.25, 0.75, 0), \qquad (4.12)$$
$$X_{21} = (0, 1, 0, 0, 0), \qquad (4.13)$$
$$Y_1 = (0.6, 0.4, 0), \qquad (4.14)$$
$$X_{12} = (0, 0, 0, 0, 0.5, 0.5), \qquad (4.15)$$
$$X_{22} = (0, 0, 0, 0.8, 0.2), \qquad (4.16)$$
$$Y_2 = (0, 0.5, 0.5). \qquad (4.17)$$

Thus, for the first data pair,

$$R_{X_{21} \times X_{11}} = \begin{bmatrix} 0 \\ 1 \\ 0 \\ 0 \\ 0 \end{bmatrix} [0\ 0\ 0\ 0.25\ 0.75\ 0] \tag{4.18}$$

$$= \begin{bmatrix} 0 & 0 & 0 & 0 & 0 & 0 \\ 0 & 0 & 0 & 0.25 & 0.75 & 0 \\ 0 & 0 & 0 & 0 & 0 & 0 \\ 0 & 0 & 0 & 0 & 0 & 0 \\ 0 & 0 & 0 & 0 & 0 & 0 \end{bmatrix} . \tag{4.19}$$

The fuzzy relations between the inputs and the output are

$$R_{y,B^1,1} = 0.6 \wedge R_{X_{21} \times X_{11}} = \begin{bmatrix} 0 & 0 & 0 & 0 & 0 & 0 \\ 0 & 0 & 0 & 0.15 & 0.45 & 0 \\ 0 & 0 & 0 & 0 & 0 & 0 \\ 0 & 0 & 0 & 0 & 0 & 0 \\ 0 & 0 & 0 & 0 & 0 & 0 \end{bmatrix} , \tag{4.20}$$

$$R_{y,B^2,1} = 0.4 \wedge R_{X_{21} \times X_{11}} = \begin{bmatrix} 0 & 0 & 0 & 0 & 0 & 0 \\ 0 & 0 & 0 & 0.1 & 0.3 & 0 \\ 0 & 0 & 0 & 0 & 0 & 0 \\ 0 & 0 & 0 & 0 & 0 & 0 \\ 0 & 0 & 0 & 0 & 0 & 0 \end{bmatrix} , \tag{4.21}$$

$$R_{y,B^3,1} = 0.0 \wedge R_{X_{21} \times X_{11}} = \begin{bmatrix} 0 & 0 & 0 & 0 & 0 & 0 \\ 0 & 0 & 0 & 0 & 0 & 0 \\ 0 & 0 & 0 & 0 & 0 & 0 \\ 0 & 0 & 0 & 0 & 0 & 0 \\ 0 & 0 & 0 & 0 & 0 & 0 \end{bmatrix} . \tag{4.22}$$

Similarly, the fuzzy relations for the second data pair can be obtained:

$$R_{X_{22} \times X_{12}} = \begin{bmatrix} 0 \\ 0 \\ 0 \\ 0.8 \\ 0.2 \end{bmatrix} [0\ 0\ 0\ 0\ 0.5\ 0.5] \tag{4.23}$$

$$= \begin{bmatrix} 0 & 0 & 0 & 0 & 0 & 0 \\ 0 & 0 & 0 & 0 & 0 & 0 \\ 0 & 0 & 0 & 0 & 0 & 0 \\ 0 & 0 & 0 & 0 & 0.4 & 0.4 \\ 0 & 0 & 0 & 0 & 0.1 & 0.1 \end{bmatrix} . \tag{4.24}$$

The fuzzy relations between the inputs and the output are

$$R_{y,B^1,2} = 0 \wedge R_{X_{22} \times X_{12}} = \begin{bmatrix} 0\,0\,0\,0\,0\,0 \\ 0\,0\,0\,0\,0\,0 \\ 0\,0\,0\,0\,0\,0 \\ 0\,0\,0\,0\,0\,0 \\ 0\,0\,0\,0\,0\,0 \end{bmatrix}, \tag{4.25}$$

$$R_{y,B^2,2} = 0.5 \wedge R_{X_{22} \times X_{12}} = \begin{bmatrix} 0\,0\,0\,0\,0 & 0 \\ 0\,0\,0\,0\,0 & 0 \\ 0\,0\,0\,0\,0 & 0 \\ 0\,0\,0\,0\,0.2 & 0.2 \\ 0\,0\,0\,0\,0.05 & 0.05 \end{bmatrix}, \tag{4.26}$$

$$R_{y,B^3,2} = 0.5 \wedge R_{X_{22} \times X_{12}} = \begin{bmatrix} 0\,0\,0\,0\,0 & 0 \\ 0\,0\,0\,0\,0 & 0 \\ 0\,0\,0\,0\,0 & 0 \\ 0\,0\,0\,0\,0.2 & 0.2 \\ 0\,0\,0\,0\,0.05 & 0.05 \end{bmatrix}. \tag{4.27}$$

To sum up the fuzzy relations of the two data pairs, we have:

$$R_{y,B^1} = \begin{bmatrix} 0\,0\,0\,0 & 0 & 0 \\ 0\,0\,0\,0.15 & 0.45 & 0 \\ 0\,0\,0\,0 & 0 & 0 \\ 0\,0\,0\,0 & 0 & 0 \\ 0\,0\,0\,0 & 0 & 0 \end{bmatrix}, \tag{4.28}$$

$$R_{y,B^2} = \begin{bmatrix} 0\,0\,0\,0 & 0 & 0 \\ 0\,0\,0\,0.1 & 0.3 & 0 \\ 0\,0\,0\,0 & 0 & 0 \\ 0\,0\,0\,0 & 0.2 & 0.2 \\ 0\,0\,0\,0 & 0.05 & 0.05 \end{bmatrix}, \tag{4.29}$$

$$R_{y,B^3} = \begin{bmatrix} 0\,0\,0\,0\,0 & 0 \\ 0\,0\,0\,0\,0 & 0 \\ 0\,0\,0\,0\,0 & 0 \\ 0\,0\,0\,0\,0.2 & 0.2 \\ 0\,0\,0\,0\,0.05 & 0.05 \end{bmatrix}. \tag{4.30}$$

The threshold λ plays a significant role in generating the rule base. If λ is set to 0.2 for all the fuzzy relations, then we have the following fuzzy relations:

$$R_{y,B^1} = \begin{bmatrix} 0\,0\,0\,0\,0 & 0 \\ 0\,0\,0\,0\,B^1 & 0 \\ 0\,0\,0\,0\,0 & 0 \\ 0\,0\,0\,0\,0 & 0 \\ 0\,0\,0\,0\,0 & 0 \end{bmatrix}, \tag{4.31}$$

$$R_{y,B^2} = \begin{bmatrix} 0\,0\,0\,0\,0 & 0 \\ 0\,0\,0\,0\,B^2 & 0 \\ 0\,0\,0\,0\,0 & 0 \\ 0\,0\,0\,0\,B^2 & B^2 \\ 0\,0\,0\,0\,0 & 0 \end{bmatrix}, \tag{4.32}$$

$$R_{y,B^3} = \begin{bmatrix} 0\,0\,0\,0\,0 & 0 \\ 0\,0\,0\,0\,0 & 0 \\ 0\,0\,0\,0\,0 & 0 \\ 0\,0\,0\,0\,B^3 & B^3 \\ 0\,0\,0\,0\,0 & 0 \end{bmatrix}. \tag{4.33}$$

However, we find that the above fuzzy relations are not consistent. For example, from R_{y,B^1}, it says that if x_1 is A_1^5 and x_2 is A_2^2, then y is B_1. However, from R_{y,B^2}, it says that if x_1 is A_1^5 and x_2 is A_2^2, then y is B_2. To resolve such conflicting fuzzy rules, the thresholds need to be adjusted. For example, set $\lambda_1 = 0.4$, $\lambda_2 = 0.4$, and $\lambda_2 = 0.15$, then we can obtain the following fuzzy relations from the two given data pair:

$$R_{y,B^3} = \begin{bmatrix} \times & \times & \times & \times & \times & \times \\ \times & \times & \times & \times & B^1 & \times \\ \times & \times & \times & \times & \times & \times \\ \times & \times & \times & \times & B^3 & B^3 \\ \times & \times & \times & \times & \times & \times \end{bmatrix}, \tag{4.34}$$

where the symbol '$\times$' means that the fuzzy rule on this position is not determined. Thus, three fuzzy rules are generated from the two pairs of data.

If x_1 is A_1^5 and x_2 is A_2^2, then y is B^1;
If x_1 is A_1^5 and x_2 is A_2^4, then y is B^3;
If x_1 is A_1^6 and x_2 is A_2^4, then y is B^3.

The advantage of this method is that the fuzzy partitions are guaranteed to be well distinguishable and complete because they are given by the designer. In addition, the number of fuzzy rules are moderate if the dimensionality of the system is not high. With the increase of the input dimensionality, the number of fuzzy rules may increase exponentially, which of course may be limited by the number of available data pairs.

However, it can be seen that with the increase of the data pairs, it will become more and more difficult to resolve conflicts between rules generated from different data pairs. One straightforward way to do this is to compare the strength of the fuzzy relation of the corresponding fuzzy rules. Besides, it can also be seen that a data pair may generate more than one fuzzy rule.

4.3 Wang-Mendel's Method

The method proposed in [236] for generating fuzzy rules has been very popular in the field fuzzy systems, not only because of its simplicity and effectiveness in data-driven fuzzy rule generation, but also because of the proof of the universal approximation capability of fuzzy systems.

The method consists of four steps:

1. Define the partition of each linguistic variable. Suppose the universe of discourse of the variable x is $[\underline{x}, \bar{x}]$, where $\underline{x}$ and $\bar{x}$ are the lower and upper bounds of x, respectively, then, this interval is divided into $2P+1$ regions, with each region being labeled with a linguistic term. The membership function for each linguistic term (a fuzzy subset) is usually a triangular function, or a Gaussian function. It should be emphasized that P should usually be less than or equal to 4, so that the total number of subsets in a fuzzy partition is not more than 10.

2. Generate fuzzy rules from the available data pairs. This step includes two sub-steps. At first, the degree of a given data to each subset of the variable is calculated. Assume the fuzzy system has n inputs and one outputs, then we have

$$DX_{j,k}^{(i)} = A_j^k(x_j^{(i)}), \tag{4.35}$$

$$DY_k^{(i)} = B^k(y^{(i)}), \tag{4.36}$$

where i denotes the i-th data pair, $j = 1, 2, ..., n$ is the j-th input, $k = 1, 2, ..., 2P+1$ is the k-th fuzzy subset in the fuzzy partition, $A_j^k(x_j)$ represents the membership function k-th fuzzy subset in the fuzzy partition of x_j, $B^k(y)$ represents the k-th membership function of output y. Consider the data pair $D^{(1)}$ in Fig. 4.1. It can be seen that $x_1^{(1)}$ has a degree of 0.75 in A_1^4, a degree of 0.25 in A_1^5 and a degree of 0 in all other subsets. Similarly, $x_2^{(1)}$ has a degree of 1 in A_2^2 and a degree of 0 in all other sets. The output has a degree of 0.6 in B_1, a degree of 0.4 in B_2 and a degree of 0 in B_3.

Then a fuzzy rule $R^{(i)}$ can be extracted from the given data pair $D^{(i)}$:

$$R^{(i)}: \text{If } x_1 \text{ is } A_1^{k_1^i}, x_2 \text{ is } A_2^{k_2^i}, \cdots, x_n \text{ is } A_n^{k_n^i}, \text{ then } y \text{ is } B^{k^i},$$

where k_j^i and k^i are determined as follows:

$$k_j^i = k^\star, \text{ if } DX_{j,k^\star}^{(i)} = \max_{k=1}^{2P+1}\{DX_{j,k}^{(i)}\}, \tag{4.37}$$

$$k^i = k^\star, \text{ if } DY_{k^\star}^{(i)} = \max_{k=1}^{2P+1}\{DY_k^{(i)}\}. \tag{4.38}$$

For the first data pair in the example in Fig. 4.1, we have

$$k_1^1 = 5, \tag{4.39}$$

$$k_2^1 = 2, \tag{4.40}$$

$$k^1 = 1. \tag{4.41}$$

Thus we have the following fuzzy rule from the first data pair:

$$R^{(1)}: \text{If } x_1 \text{ is } A_1^5, x_2 \text{ is } A_2^2, \text{ then } y \text{ is } B^1. \qquad (4.42)$$

It happens that a given data has an equal degree in two different fuzzy subset. For example, $x_1^{(2)}$ has a degree of 0.5 in both A_1^5 and A_1^6, $y^{(2)}$ has a degree of 0.5 in both B^2 and B^0. In this case, it should be determined by the designer which fuzzy subset should be taken in the fuzzy rule. Of course, the fuzzy rule can also take the following form:

$$\text{If } x_1 \text{ is } A_1^5 \text{ or } A_1^6, x_2 \text{ is } A_2^4, \text{ then } y \text{ is } B^2 \text{ or } B^3.$$

Nevertheless, such rules should be generally avoided because they have negative influence on the interpretability of fuzzy rules. In our case, we adopt the following fuzzy rule for the second data pair:

$$R^{(2)}: \text{If } x_1 \text{ is } A_1^6, x_2 \text{ is } A_2^4, \text{ then } y \text{ is } B^2. \qquad (4.43)$$

3. Remove conflicting rules. It is not unusual that conflicting rules may be generated in this data-driven fuzzy rule generation process. Two rules are considered to be inconsistent if they have the same condition part but different action part. The conflict is resolved by assigning a truth degree to each generated fuzzy rule. The truth degree consists of two parts, one is the combined degree of the given data pair to the generated fuzzy rule, the other is the user's belief on the given data pair:

$$DR^{(i)} = \prod_{j=1}^{n} A_j^{k_j^i} \cdot B^{k^i} \cdot d^{(i)}, \qquad (4.44)$$

where $d^{(i)}$ is the belief of the user on data $D^{(i)}$. Then, if more than one fuzzy rule has the same condition part but different action part, only the rule with the maximal truth degree is kept.

4. Determine the output from the generated fuzzy rules. The output of the whole fuzzy system is determined as follows:

$$y = \frac{\sum_{i=1}^{N} w^i y_c^i}{\sum_{i=1}^{N} w^i}, \qquad (4.45)$$

$$w^i = \prod_{j=1}^{n} A_j^{k_j^i}(x_j), \qquad (4.46)$$

where N is the number of generated fuzzy rules in the rules base, y_c^i is the center of the fuzzy set B^{k^i}.

One problem in this method is that the generated fuzzy rules are not able to reproduce the output for a given input. For example, from the two fuzzy rules in equations (4.42) and (4.43), and given the inputs $x_1 = 0.75$, and

$x_2 = 0.25$, we get $y = 0$ instead of $y = 0.2$. For the given inputs $x_1 = 0.9$ and $x_2 = 0.8$, the output generated from the two fuzzy rules is 0.2 instead of 0.75. It is found that the smaller the truth degree, the larger the error will be. It can easily be calculated that the truth degree of the two rules is 0.45 and 0.2, respectively, assuming that the user's belief on the data pairs is 1.

4.4 A Direct Method

In [114], a method for generating fuzzy rules from data is suggested. This method has shown to be very effective for high-dimensional systems.

The main weakness of the Wang-Mendel's method for fuzzy rule generation from data is that the structure of fuzzy rules needs to be predefined. This kind of predefined fuzzy partitions may not be optimal. In addition, if the input dimension of the system is high and the number of data is large, the number of fuzzy rules becomes huge.

Optimal rules can reduce the number of fuzzy rules needed to approximate a function. As pointed out in [142], optimal fuzzy rules cover extrema. This is a very important hint for data-driven fuzzy rule generation. In the following, a method for fuzzy rule generation that takes advantage of this property will be introduced.

The procedure of fuzzy rule generation can be carried out in the following steps.

1. Divide the given data into a certain number of patches. Suppose the total number of given data sets is N. If we divide the whole data sets into M patches, the number of data sets in each patch will be $P = N/M$, assuming that P is an integer. The patch size can be adjusted properly based on the smoothness of the output data. Normally, when the output varies rapidly, the patch size should be smaller, on the other hand, the patch size can be a little larger if the output surface is smooth.

2. In each patch of the data, find the two data pairs where the output reaches the minimum and maximum, respectively, refer to Fig. 4.3. Suppose for patch j, the two data pairs are:

$$(x_1^{(i)}, x_2^{(i)}, \cdots, x_n^{(i)}, y^{(i)});$$
$$(x_1^{(j)}, x_2^{(j)}, \cdots, x_n^{(j)}, y^{(j)}),$$

where, n is the dimension of the input, and

$$y^{(i)} = \max_{k=1}^{P} y^{(k)}, \tag{4.47}$$
$$y^{(j)} = \min_{k=1}^{P} y^{(k)}. \tag{4.48}$$

This reduces the number of fuzzy rules significantly when the number of available data is huge.

3. For these two data pairs, two TSK fuzzy rules can be generated:

R_{j1}: If x_1 is $A_{j1}^{max}(x_1)$, ... , x_n is $A_{jn}^{max}(x_n)$, then y is y_j^{max};
R_{j2}: If x_1 is $A_{j1}^{min}(x_1)$, ... , x_n is $A_{jn}^{min}(x_n)$, then y is y_j^{min},

where, A_{ji} is a fuzzy set whose membership function is directly determined by the given data pair. If the Gaussian function is used for membership function, we have:

$$A_{ji}^{max}(x_i) = \exp(-(x_i - x_{ji}^{max})^2/\sigma_{x_i}^{max}), \qquad (4.49)$$

$$A_{ji}^{min}(x_i) = \exp(-(x_i - x_{ji}^{min})^2/\sigma_{x_i}^{min}), \qquad (4.50)$$

where x_{ji}^{max} and x_{ji}^{min} are termed as the center of the Gaussian function, $\sigma_{x_i}^{max}$ and $\sigma_{x_i}^{min}$ are known as the width of the Gaussian function. The width of the Gaussian functions can be determined according to the universe of discourse of the variable, taking into account the fuzziness of the fuzzy sets.

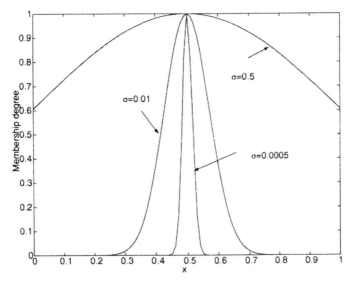

Fig. 4.2. Gaussian fuzzy membership functions.

The value of δ in the Gaussian fuzzy membership functions can be seen as a linguistic hedge. That is, whereas the center of the Gaussian function mainly determines the linguistic term, say "small", "medium" and "large", δ is the linguistic hedge that modifies the linguistic term. Fig. 4.2 shows the shape of different Gaussian fuzzy membership functions when σ changes. If the universe of discourse is normalized to $[0, 1]$, it is recommended that $0.0005 \leq \sigma \leq 0.5$. It is seen that if σ equals 0.01, the shape of the fuzzy membership function looks like very "typical". If σ is either much larger than 0.5 or much smaller than 0.0005, the fuzzy set becomes non-fuzzy.

In case triangular functions are used, the membership function can be determined as follows:

$$A_{ji}^{max}(x_i) = 1 - \frac{2|x_i - x_{ji}^{max}|}{b_{ji}^{max}}, \tag{4.51}$$

$$A_{ji}^{min}(x_i) = 1 - \frac{2|x_i - x_{ji}^{min}|}{b_{ji}^{min}}. \tag{4.52}$$

Similarly, b_{ji} is the width of the fuzzy set to be determined properly.

In this way, $2P$ fuzzy rules will be generated in total, which will be much smaller than N, because usually, the patch size could be 10 or larger. Fig. 4.3 illustrates the minimum and the maximum, A and B on patch j, from each of which one fuzzy rule will be generated.

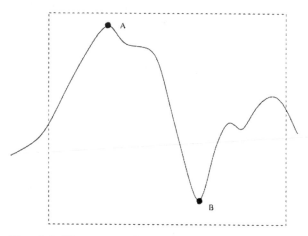

Fig. 4.3. Illustration of a data patch and its extrema.

4. Check the similarity of the currently generated rules to the existing rules. Usually, there would be many redundant or sometimes inconsistent rules if the generated fuzzy rules are not properly checked. Redundant or inconsistent rules are generated when the inputs of different rules are quite similar. If the consequent parts of these rules are also similar, the rules may be redundant. If, usually, due to noises in the data, the consequent parts are significantly different, inconsistent rules will be generated. To prevent the algorithm from generating fuzzy rules that are redundant or inconsistent, a similarity measure for fuzzy rules needs to be developed. In this work we use a measure called similarity of rule premise (SRP) proposed in [126]. Consider the following two fuzzy rules,

R_i: If x_1 is $A_{i1}(x_1)$, x_2 is $A_{i2}(x_2)$, ..., x_n is $A_{in}(x_n)$, then y is $B_i(y)$;
R_k: If x_1 is $A_{k1}(x_1)$, x_2 is $A_{k2}(x_2)$, ..., x_n is $A_{kn}(x_n)$, then y is $B_k(y)$;

the SRP of the two rules is defined by:

$$SRP(i, k) = \min_{j-1}^{n} S(A_{ij}, A_{kj}), \tag{4.53}$$

where $S(A, B)$ is a fuzzy similarity measure for fuzzy sets A and B, which is defined by:

$$S(A, B) = \frac{M(A \cap B)}{M(A) + M(B) - M(A \cap B)}, \tag{4.54}$$

where $M(A)$ is the size of fuzzy set A:

$$M(A) = \int_{-\infty}^{+\infty} A(x)dx. \tag{4.55}$$

Refer to [126] for the details for computing the fuzzy similarity measure. By checking the SRP of the fuzzy rules, redundant and inconsistent rules can be removed. In this way, the rule base can be simplified greatly.

Consider again the two data pairs discussed in Section 1.2. From the two data pairs, the following two fuzzy rules can be obtained:

$$R^{(1)}: \text{If } x_1 \text{ is } A_1^1, x_2 \text{ is } A_2^1, \text{ Then } y \text{ is } B^1, \tag{4.56}$$

$$R^{(2)}: \text{If } x_1 \text{ is } A_1^2, x_2 \text{ is } A_2^2, \text{ Then } y \text{ is } B^2, \tag{4.57}$$

assuming triangular membership functions are adopted with a width of 0.4. The membership functions are shown in Fig. 4.4. It is noticed that if the center value of a fuzzy set is at one end of the universe of discourse, the shape of the membership function (A_1^2) should be modified a little to satisfy the interpretability conditions.

4.5 An Adaptive Fuzzy Optimal Controller

Consider a class of nonlinear systems:

$$x(k + 1) = f(x(k), \cdots, x(k - P_1), u(k), \cdots, u(k - P_2)), \tag{4.58}$$

where $f(\cdot)$ is an unknown nonlinear system, $x(k), \cdots, x(k-P_1)$ and $u(k), \cdots,$ $u(k - P_2)$ are system states and control, P_1 and P_2 are structure parameters of the system. The fuzzy rules to be generated for control are in the following form:

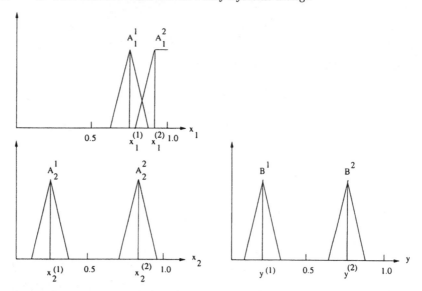

Fig. 4.4. Fuzzy membership functions obtained directly from data pairs.

$$R^1 : \text{If } e(k) \text{ is } A^1, \ \Delta e(k) \text{ is } B^1, \text{ then } u(k) \text{ is } C^1; \tag{4.59}$$

$$R^2 : \text{If } e(k) \text{ is } A^2, \ \Delta e(k) \text{ is } B^2, \text{ then } u(k) \text{ is } C^2; \tag{4.60}$$

$$\vdots \qquad\qquad\qquad \vdots \tag{4.61}$$

$$R^N : \text{If } e(k) \text{ is } A^N, \ \Delta e(k) \text{ is } B^N, \text{ then } u(k) \text{ is } C^N. \tag{4.62}$$

In the fuzzy rules, $e(k) = x^d(k) - x(k)$, $x^d(k)$ is the desired state of the system at time k, $\Delta e(k) = (e(k) - e(k-1))/T$, where T is the sampling period, and $u(k)$ is the control input. $A^i, B^i, C^i, i = 1, ..., N$ are fuzzy sets defined on the universe of discourse of $e(k)$, $\Delta e(k)$ and $u(k)$. The output of the entire fuzzy controller can be written as:

$$u(k) = \sum_{i=1}^{N} \frac{w^i z^i}{\sum_{i=1}^{N} w^i}, \tag{4.63}$$

where,

$$w^i = \mu_{A^i}(e(k)) \cdot \mu_{B^i}(\Delta e(k)), \tag{4.64}$$

$$z^i = \mu_{C^i}^{-1}(1), \tag{4.65}$$

where z^i is obviously the center of the fuzzy set C^i.

The parameters of the fuzzy rules need to be determined. For the fuzzy membership function, it can be set up heuristically. One advantage of doing this is that the interpretability of the fuzzy systems will be preserved and the learning algorithm can significantly simplified. Define the following quadratic error function that is widely used in optimal control:

$$J = \frac{1}{2}\{[x^d(k+1) - x(k+1)]^T Q[x^d(k+1) - x(k+1)] + u^T(k)Ru(k)\},$$

$$(4.66)$$

where Q and R are symmetric positive definite weighting matrices.

The learning algorithm for the center (z^i) of the fuzzy sets C^i can be derived as follows:

$$\Delta z^i = -\eta \frac{\partial J}{\partial z^i}, \qquad (4.67)$$

where,

$$\frac{\partial J}{\partial z^i} = \frac{1}{2} \frac{\partial [x^d(k+1) - x(k+1)]^T Q[x^d(k+1) - x(k+1)]}{\partial z^i}$$

$$+ \frac{1}{2} \frac{\partial [u^T(k)Ru(k)]}{\partial z^i} \qquad (4.68)$$

$$= -Q[x^d(k+1) - x(k+1)] \frac{\partial x(k+1)}{\partial u(k)} \frac{\partial u(k)}{\partial z^i} + Ru(k) \frac{\partial u(k)}{\partial z^i} \quad (4.69)$$

$$= [Ru(k) - Qe(k+1) \frac{\partial x(k+1)}{\partial u(k)}] \frac{\partial u(k)}{\partial z^i}. \qquad (4.70)$$

According to Equation (4.63), we have

$$\frac{\partial u(k)}{\partial z^i} = \frac{w^i}{\sum_{i=1}^{N} w^i}. \qquad (4.71)$$

The partial derivative of the system state $x(k+1)$ with regard to the system input $u(k)$ can be derived if the model of system (4.58) is known. For example, if the controlled plant is a class of affine nonlinear systems:

$$x(k+1) = \alpha(x(k)) + \beta(x(k))u(k), \qquad (4.72)$$

where $\alpha(\cdot)$ and $\beta(\cdot)$ are nonlinear functions, then the learning algorithm can be expressed as

$$\Delta z^i = -\eta[Ru(k) - Qe(k+1)\beta(x(k))] \frac{w^i}{\sum_{i=1}^{N} w^i}. \qquad (4.73)$$

If the system dynamics is unknown, then we need to estimate the derivative:

$$\frac{\partial x(k+1)}{\partial u(k)} \approx \frac{x(k+1) - x(k)}{u(k) - u(k-1)}. \qquad (4.74)$$

To simplify it further, the estimated derivative can be replaced by the sign function:

$$\text{sgn} \left\{ \frac{x(k+1) - x(k)}{u(k) - u(k-1)} \right\}. \qquad (4.75)$$

It is interesting to notice that the sign function may be determined from heuristics and experience about the system. Actually, it is found better to use a fuzzy system to estimate the derivative [116], which is less sensitive to noise. Besides, it will improve the performance significantly if the control system is multivariable.

To avoid computational difficulties, for example, $u(k)$ equals $u(k-1)$, the learning algorithm for the consequence parameters can be rewritten as:

$$\Delta z^i = -\eta \{[Ru(k) - Qe(k+1)\mathrm{sgn}[(x(k+1) - x(k)) \cdot [u(k) - u(k-1)]]\} w^i / \sum_{i=1}^{N} w^i. \tag{4.76}$$

Usually, it is insufficient to adapt the consequence parameters only, i.e., the parameters of the membership functions in the rule premise should also be updated. Suppose the Gaussian functions in the following form are used for fuzzy membership functions:

$$\mu_{A^i}(e(k)) = \exp[-b_e^i \cdot (e(k) - a_e^i)^2], \tag{4.77}$$

$$\mu_{B^i}(\Delta e(k)) = \exp[-b_{\Delta e}^i \cdot (\Delta e(k) - a_{\Delta e}^i)^2], \tag{4.78}$$

then the learning algorithm for membership parameters can be derived as follows:

$$\Delta a_e^i = -2\xi[Ru(k) - Qe\frac{\partial x(k+1)}{\partial u(k)}][z^i - u(k)]b_e^i(e - a_e^i)\bar{w}^i, \tag{4.79}$$

$$\Delta b_e^i = \xi[Ru(k) - Qe\frac{\partial x(k+1)}{\partial u(k)}][z^i - u(k)](e - a_e^i)^2\bar{w}^i, \tag{4.80}$$

$$\Delta a_{\Delta e}^i = -2\xi[Ru(k) - Qe\frac{\partial x(k+1)}{\partial u(k)}][z^i - u(k)]b_e^i(\Delta e - a_{\Delta e}^i)\bar{w}^i, \tag{4.81}$$

$$\Delta b_{\Delta e}^i = \xi[Ru(k) - Qe\frac{\partial x(k+1)}{\partial u(k)}][z^i - u(k)](\Delta e - a_{\Delta e}^i)^2\bar{w}^i, \tag{4.82}$$

where the time index for the system response error $(e(k+1))$ and the change of error $(\Delta e(k+1))$ is omitted, and

$$\bar{w}^i = \frac{w^i}{\sum_{i=1}^{N} w^i}. \tag{4.83}$$

As previously discussed, an estimate of the system derivative is necessary if the system is unknown.

To illustrate the effectiveness of the method, a fuzzy controller will be designed for the following nonlinear system:

$$x(k+1) = \frac{x(k)x(k-1)}{1. + x^2(k) + x^2(k-1)} + u(k-1) + 0.5u(k). \tag{4.84}$$

Let $Q = 0.6I$, $R = 0.01I$, where I is the unit matrix, and the linguistic variable for the error E is defined as follows:

$$E = \{(-2,1)/NB, (-1,1)/NM, (0,1)/ZO, (1,1)/PM, (2,1)/PB\},$$

where the two values in the parenthesis denote the center and the width of the corresponding membership function. The linguistic variable for the error change ΔE is defined as follows:

$$\Delta E = \{(-1,1)/N, (0,1)/ZO, (1,1)/P\}.$$

All consequent parameters are initialized to zero. The learning results of the membership functions for the error and change of error are shown in Fig. 4.5, the desired and the real system response are given in Fig. 4.6.

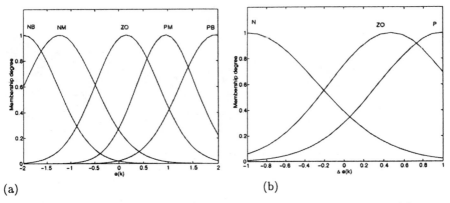

(a)   (b)

Fig. 4.5. Membership functions after learning for (a) $e(k)$, (b) $\Delta e(k)$.

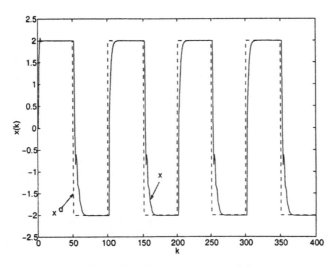

Fig. 4.6. Desired and real response of the system.

The trained fuzzy controller exhibits a quite good performance for a sinus input, see Fig. 4.7.

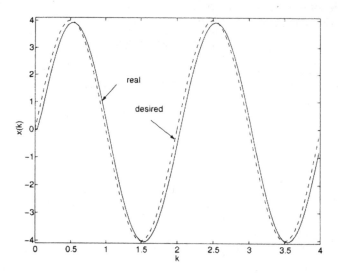

Fig. 4.7. Sinus response of the system.

4.6 Summary

In this chapter, three methods for generating fuzzy rules directly from data have been presented. In the fuzzy inference method and the Wang-Mendel's method, the structure of the fuzzy system is heuristically predefined. An advantage of the predefined rule structure is that the interpretability of the fuzzy system is generally good, provided that the number of generated fuzzy rules is moderate. However, the main disadvantage of these methods is that a predefined fuzzy rule structure is hardly optimal. Furthermore, the number of fuzzy rules may become huge if the input dimension of the system is high.

In the direct method, the structure of the rule system is determined by the given data. Fuzzy rules are generated on extrema of the output, which has been proven to be optimal. To reduce the number of fuzzy rules, similarity of the rules is checked to ensure that rules that are similar to or inconsistent with existing rules are not generated. One weakness of this method is that a large number of fuzzy subsets in a fuzzy partition may be generated. Therefore, interpretability checking and improvement may be necessary.

Finally, a method for generating adaptive fuzzy optimal controller is presented. This method does not require any *a priori* knowledge, and is fully data-driven. With the help of the gradient method, the fuzzy controller can be adapted on-line.

5. Neural Network Based Fuzzy Systems Design

5.1 Neurofuzzy Systems

Fuzzy rules are able to represent knowledge that is understandable to human beings. Traditional fuzzy rules are usually generated from expert knowledge and human heuristics. This gives rise to two main drawbacks of traditional fuzzy systems for modeling and control. First, the fuzzy rules are very simple and the performance of the fuzzy system is low. In most cases, fuzzy memberships are determined heuristically and therefore, the knowledge represented by the fuzzy rules may be shallow. Second, it is difficult to efficiently extract fuzzy rules for high-dimensional systems due to the limitation of human thinking. In particular, traditional fuzzy systems are lack of learning capability, whereas learning is one of the most important features of intelligent systems.

The marriage of neural networks with fuzzy systems provides an excellent example of combining two new techniques. It is well known that neural networks are capable of learning on the basis of a set of training data. Therefore, if a fuzzy system can be modeled by a neural network model, existing sophisticated learning algorithms developed in the field of neural networks can be employed to adapt fuzzy systems so that a fuzzy system is able to learn. Therefore, a neurofuzzy system is a fuzzy system that can learn. Two main approaches to the construction of a neurofuzzy system have been developed. In the first approach, a neural network learning algorithm is directly applied to a fuzzy system. For example, in [15], the reinforcement learning is employed to adjust the parameters of the fuzzy rules, without representing the fuzzy system in a neural network structure. In the second approach, which is also the most popular in the neurofuzzy research, a fuzzy system is first interpreted with a general or specific neural network architecture. After that, learning algorithms developed for training neural networks can be used to train the fuzzy system. In [158], a fuzzy rule system is represented by a five-layer neural network architecture (see Fig. 5.1). The most well known neurofuzzy system is the adaptive-network-based fuzzy inference system (AN-FIS), as shown in Fig. 5.2, which is proposed in [108, 109]. Similar neurofuzzy architectures have also been proposed in [235].

In a more general sense, fuzzy systems and neural networks can be described by a unified framework that is called adaptive networks [109]. Radial-

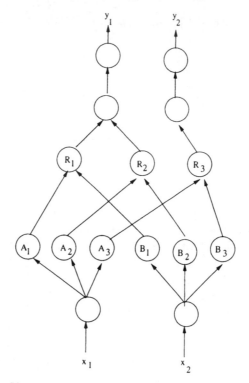

Fig. 5.1. A five layer neurofuzzy structure.

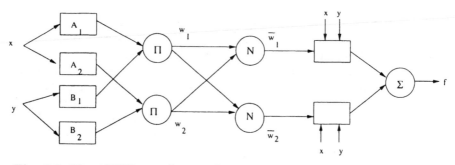

Fig. 5.2. The ANFIS neurofuzzy architecture. 'N' means weigh normalization.

basis-function networks (RBFN) and multilayer perceptrons (MLP) are two
most widely used feedforward neural networks. Among them, the RBFNs has
been found to be mathematically equivalent to a class of fuzzy systems [110].
Meanwhile, it has also been shown that MLPs can be interpreted as fuzzy
rules [31].

The unique property of neurofuzzy systems is that neurofuzzy systems
have the features of both fuzzy systems and neural works. In other words, a
neurofuzzy system is not only interpretable like fuzzy systems, but is also able

to learn like neural networks. Therefore, the preservation of interpretability of neurofuzzy systems is a very important issue in the research of neurofuzzy systems. Methods for the preservation of interpretability of fuzzy systems can be classified into the following categories:

• Constraints on the parameters. The simplest way to preserve the predefined linguistic meaning of the fuzzy sets during learning is to impose constraints on the parameters of the fuzzy membership functions. A straightforward way of achieving this is to reduce the number of free parameters in defining fuzzy subsets. For example, in Fig. 5.3(a), constraints are imposed on the parameters of the fuzzy membership functions. In contrast, the fuzzy membership functions in Fig. 5.3(b) have more free parameters. Therefore, it is generally much easier in case (a) to preserve interpretability of fuzzy systems than in case (b).

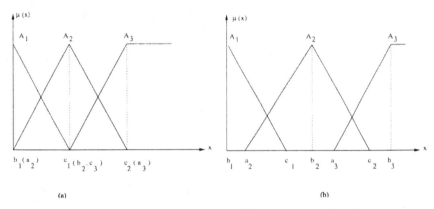

(a)   (b)

Fig. 5.3. Constraints on parameters of fuzzy membership functions.

• Constrains on the architecture. To prevent the neurofuzzy model from losing the predefined physical meanings, constraints can be imposed on the model architecture. For example, in [174], it is required that some connections in both input layer and output layer use the same weight, as illustrated in Fig. 5.4 for a two input, single output neurofuzzy system with four rules.
• Merging of similar membership functions. By merging similar membership functions [205], it is able to improve the distinguishability of fuzzy partitions and to reduce the number of fuzzy subsets. Both will make contribution to the improvement of interpretability of neurofuzzy systems.
• Regularization. In most neurofuzzy systems, the learning algorithm is aimed at reducing the approximation error only. In [114], distinguishability constraints are incorporated in learning to maintain the interpretability of neurofuzzy systems.

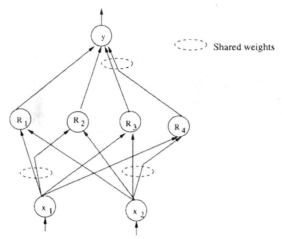

Fig. 5.4. The NEFPROX neurofuzzy architecture.

In this chapter, a neurofuzzy model based on a hybrid neural network model is presented. The main advantage of the neurofuzzy model is that higher order TSK fuzzy systems can be implemented. Unlike the ANFIS model, minimization as well as multiplication can be used as the T-norm operator. The neurofuzzy model is applied to modeling and control of complex systems to show its effectiveness.

5.2 The Pi-sigma Neurofuzzy Model

5.2.1 The Takagi-Sugeno-Kang Fuzzy Model

Takagi-Sugeno-Kang (TSK) fuzzy model is one of the two most widely use fuzzy models for modeling and control. The main difference between these two fuzzy models is that the consequent part of the TSK model is a linear or nonlinear function of the input variables instead of a fuzzy subset. In the recent years, the TSK fuzzy model has gained more and more attention, especially in fuzzy control. This is due to the fact that the TSK fuzzy model has stronger and more flexible representation capability than the Mamdani fuzzy model.

Consider the i-th fuzzy rule of a first-order TSK model with n input variables:

R^i: If x_1 is A_1^i, and ... and x_n is A_n^i, then $y^i = p_0^i + \sum_{j=1}^{n} p_j^i x_j$,

where $i = 1, ..., N$ is the i-th rule, N the number of fuzzy rules in the model, $A_j^i(x_j), j = 1, ..., n$ are fuzzy subsets defined for the input variables. According to [219], the final output of the fuzzy model can be expressed by:

$$y = \frac{\sum_{i=1}^{N}(w^i y^i)}{\sum_{i=1}^{N} w^i}, \tag{5.1}$$

$$w^i = T_{j=1}^n A_j^i(x_j), \tag{5.2}$$

where T is a T norm. In this model, we take the minimum operation as the T-norm.

Given a set of experimental data, it is possible to estimate the consequent parameters in the fuzzy rules using the least square method. However, estimation of the parameters in the fuzzy membership functions is a little complicated, even if piece-wise membership functions are used.

5.2.2 The Hybrid Neural Network Model

In most neurofuzzy models, multiplication has been used as the T-norm. The main reason is that a majority of neural network learning algorithms use the gradient-based methods for parameter optimization, which requires the mathematical model to be derivable. In this chapter, the minimum operation is used for fuzzy intersection, which is one of the most important T-norm operator in fuzzy logic. Therefore, the neural network model used to represent the fuzzy system should contain not only sum and multiplication neurons, but also fuzzy neurons that are able to perform the minimum operation. To this end, the pi-sigma neural network suggested in [52] is employed with a minor extension. As an example, the neural network architecture with two inputs and one output is illustrated in Fig. 5.5, where $\sum$ denotes sum operation, $\prod$ denotes the multiplication and $\bigwedge$ means the minimum. In the figure, neurons denoted with a circle represent mathematical operations such as sum, multiplication and minimum, whereas those denoted with rectangles stand for a nonlinear function. Therefore, $A_i^j, i = 1, 2, j = 1, 2, 3$ are fuzzy membership functions, and S is the sigmoid function used in multilayer perceptrons. In the figure, H and N represent the number of nodes in the corresponding layer.

Suppose the neural network has n inputs and one output. Subnetwork 1 in Fig. 5.5 has H hidden nodes, and subnetwork 2 contains N hidden nodes, then the output of the pi-sigma network can be described as follows:

$$y = \frac{\sum_{i=1}^{N}(w^i y^i)}{\sum_{i=1}^{N} w^i}, \tag{5.3}$$

$$y^i = p_i^{(0)} + \sum_{k=1}^{H}\left(p_{ki}^{(2)} \cdot f\left(\sum_{j=1}^{n} p_{jk}^{(1)} x_j \right) \right), \tag{5.4}$$

$$w^i = \min_{j=1}^{n}\{A_j^i(x_j)\}, \tag{5.5}$$

where $f(\cdot)$ is the sigmoid function:

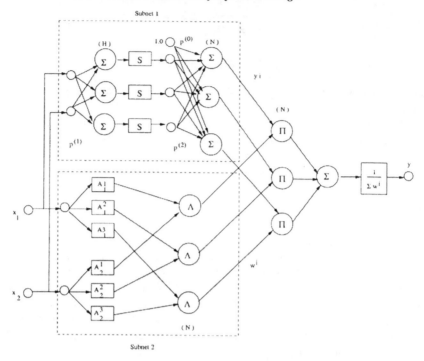

Fig. 5.5. The Pi-sigma neurofuzzy architecture.

$$f(z) = \frac{1}{1 + \exp(-z)}. \tag{5.6}$$

If we compare the hybrid neural network with the TSK fuzzy model in equations (5.1) and (5.2), it is obvious to notice that they are functionally equivalent. The only difference is that in the TSK fuzzy model, the rule consequents are usually a linear function of the input variables. In the pi-sigma neural network, it is a nonlinear function of the input variables, which is described by a feedforward neural network. Obviously, the pi-sigma neural network represents a neurofuzzy system.

5.2.3 Training Algorithms

Both the supervised learning and the reinforcement learning algorithms can be employed to train neurofuzzy systems. In this model, the supervised learning algorithm based on the gradient method will be applied. Note that the minimum operation is involved in the model, which is not derivable. Therefore, the gradient method cannot directly be applied. To address this problem, two measures can be taken. The first method makes an equivalent transformation of the minimum operator so that it can be treated as differential and thus the gradient method can be applied. The second method uses a one-step backward search algorithm to avoid the derivation operation on the minimum. Consider the minimum operation with n elements:

$$w_i = \min_{j=1}^{n}\{A_j^i(x_j)\}. \tag{5.7}$$

It can be found that the equation (5.7) is equivalent to:

$$w^i = \sum_{j=1}^{n} \prod_{m=1,m\neq j}^{n} \bigcup \left(A_m^i(x_m) - A_j^i(x_j)\right), \tag{5.8}$$

where

$$\bigcup \left(A_m^i(x_m) - A_j^i(x_j)\right) = \begin{cases} 1, & A_m^i(x_m) > A_j^i(x_j), \\ 0, & A_m^i(x_m) \leq A_j^i(x_j). \end{cases} \tag{5.9}$$

It can be seen that equation (5.8) is now derivable. Let y^d be the desired output of the neurofuzzy model, the following quadratic cost function can be defined:

$$E = \frac{1}{2}(y - y^d)^2, \tag{5.10}$$

where y is the real model output for the given input. According to the gradient method, we have

$$\begin{aligned} \frac{\partial E}{\partial p_i^{(0)}} &= \frac{\partial E}{\partial y}\frac{\partial y}{\partial y^i}\frac{\partial y^i}{\partial p_i^{(0)}} = \frac{(y-y^d)w^i}{\sum_{i=1}^{N} w^i}, \\ \frac{\partial E}{\partial p_{jk}^{(1)}} &= \frac{\partial E}{\partial y}\frac{\partial y}{\partial y^i}\frac{\partial y^i}{\partial p_{jk}^{(1)}} = \frac{(y-y^d)w^i \sum_{k=1}^{H}\left\{p_{ki}^{(2)} f'(\cdot)x_j\right\}}{\sum_{i=1}^{N} w^i}, \\ \frac{\partial E}{\partial p_{ki}^{(2)}} &= \frac{\partial E}{\partial y}\frac{\partial y}{\partial y^i}\frac{\partial y^i}{\partial p_{ki}^{(2)}} = \frac{(y-y^d)w^i f(\cdot)}{\sum_{i=1}^{N} w^i}. \end{aligned} \tag{5.11}$$

Thus, the learning algorithm for the consequent parameters can be expressed by:

$$\Delta p_i^{(0)} = -\eta \delta_i^{(1)}, \tag{5.12}$$

$$\Delta p_{jk}^{(1)} = -\eta \sum_{k=1}^{H} \left\{p_{ki}^{(2)} f'(\cdot)x_j\right\} \delta_i^{(1)}, \tag{5.13}$$

$$\Delta p_{ki}^{(2)} = -\eta f(\cdot)\delta_i^{(1)}, \tag{5.14}$$

where η is a positive learning rate and $\delta_i^{(1)}$ is the generalized error defined by

$$\delta_i^{(1)} = \frac{(y - y^d)w^i}{\sum_{i=1}^{N} w^i}. \tag{5.15}$$

The derivation of the learning algorithm for the parameters of the fuzzy membership functions is not as straightforward as that of the consequent parameters. Suppose that the fuzzy membership functions are Gaussian functions in the following form:

$$A_j^i(x_j) = \exp\left(-\left(x_j - a_j^i\right)^2 / b_j^i\right).\tag{5.16}$$

Usually, a_j^i is called the center of fuzzy subset, and b_j^i the width. Then we have

$$\frac{\partial E}{\partial a_j^i} = \frac{\partial E}{\partial y}\frac{\partial y}{\partial w^i}\frac{\partial w^i}{\partial A_j^i(x_j)}\frac{\partial A_j^i(x_j)}{\partial a_j^i}.\tag{5.17}$$

The derivation of the Gaussian function with respect to the center a_j^i is as follows:

$$\frac{\partial A_j^i}{\partial a_j^i} = \frac{2A_j^i(x_j)(x_j - a_j^i)}{b_j^i}.\tag{5.18}$$

According to the equivalent transformation of the minimum operation in equations (5.8) and (5.9), we get

$$\frac{\partial w^i}{\partial A_j^i(x_j)} = \frac{\partial\left\{\sum_{j=1}^n A_j^i(x_j)\prod_{m\neq j}\bigcup\left[A_m^i(x_m) - A_j^i(x_j)\right]\right\}}{\partial A_j^i(x_j)}$$

$$= \bigcup_{m\neq j}\left[A_m^i(x_m) - A_j^i(x_j)\right]$$

$$= \begin{cases} 1, & \text{if } A_j^i(x_j) \text{ is minimum,} \\ 0, & \text{else.} \end{cases}\tag{5.19}$$

Let

$$\delta_i^{(2)} = \frac{(y - y^d)(y^i - y)}{\sum_{i=1}^N w^i},\tag{5.20}$$

then the learning algorithm for the center of the membership functions is as follows:

$$\Delta a_j^i = \begin{cases} -2\xi(x_j - a_j^i)w^i\delta_i^{(2)}/b_j^i, & \text{if } A_j^i(x_j) \text{ is minimum,} \\ 0, & \text{else,} \end{cases}\tag{5.21}$$

where ξ is positive learning rate. Similarly, the learning algorithm for b_j^i can be obtained as follows:

$$\Delta b_j^i = \begin{cases} -\xi w^i(x_j - a_j^i)\delta_i^{(2)}/b_j^i, & \text{if } A_j^i(x_j) \text{ is minimum,} \\ 0, & \text{else.} \end{cases}\tag{5.22}$$

In fuzzy modeling and control, triangular membership functions are also widely used. Suppose an isosceles triangular membership function as shown in Fig. 5.6 is used for all membership functions, which can be described by

$$A_j^i(x_j) = 1 - \frac{2|x_j - a_j^i|}{b_j^i},\tag{5.23}$$

where a_j^i and b_j^i are the center and the width of the fuzzy subset. Thus, the

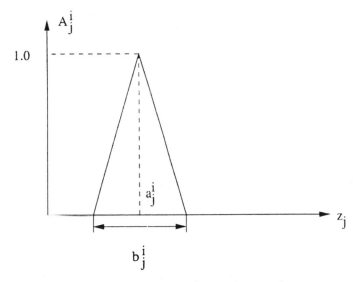

Fig. 5.6. An isosceles triangular membership function.

derivation on the triangular membership function with respect to a_j^i is

$$\frac{\partial A_j^i(x_j)}{\partial a_j^i} = 2\mathrm{sgn}(x_j - a_j^i)/b_j^i, \tag{5.24}$$

and the derivation with respect to b_j^i can be expressed as

$$\frac{\partial A_j^i(x_j)}{\partial b_j^i} = (1 - A_j^i(x_j))/b_j^i. \tag{5.25}$$

Therefore, the learning algorithm for a_j^i and b_j^i of an isosceles triangular membership function is:

$$\Delta a_j^i = \begin{cases} -2\xi\mathrm{sgn}(x_j - a_j^i)\delta_i^{(2)}/b_j^i, & \text{if } A_j^i(x_j) \text{ is minimum,} \\ 0, & \text{if } A_j^i(x_j) \text{ else,} \end{cases} \tag{5.26}$$

$$\Delta b_j^i = \begin{cases} -\xi 1.0 - A_j^i(x_j))\delta_i^{(2)}/b_j^i, & \text{if } A_j^i(x_j) \text{ is minimum,} \\ 0, & \text{if } A_j^i(x_j) \text{ else.} \end{cases} \tag{5.27}$$

In the above algorithm, the minimum operation is equivalently transformed into a new expression so that the gradient method can be used directly. If we take a closer look at the method, it is easy to find that this transformation leads the gradient search toward the input that has the minimum value. However, this is not necessarily optimal. To expound this, let us consider a fuzzy neuron that implements the minimum operation shown in Fig. 5.7. The output of the neuron is:

$$w^i = \min\{A_1^i(x_1), A_2^i(x_2), ..., A_n^i(x_n)\}. \qquad (5.28)$$

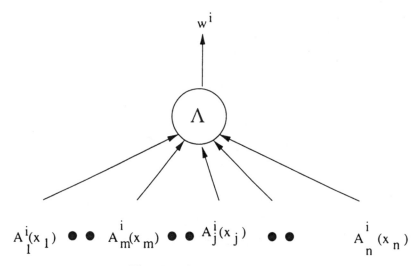

Fig. 5.7. A minimum neuron.

Suppose $A_j^i(x_j), 1 \leq j \leq n$ is the minimal element of a given input vector. Based on the previous learning algorithm, the parameters of the membership $A_j^i(x_j)$ will be updated. However, the output error of the neurofuzzy system may mainly be caused by the membership $A_m^i(x_m), 1 \leq m \leq n, m \neq j$. To address this problem, one step backward search has been introduced. The basic idea is to update the parameters of the membership functions one by one and see which update results in the maximal error reduction. For example, suppose the approximation error before learning is e_0. Then update the parameters of j-th membership function $A_j^i(x_j), j = 1, 2, ..., n$ and the new approximation error is e_j. If $e_m, 1 \leq m \leq n$ is the minimal error, it means that the update of the m-th membership function reduces the model error most significantly. Therefore, only the update of the m-th membership function is accepted and parameters of all other membership functions will keep unchanged.

5.2.4 Interpretability Issues

With the help of the learning algorithms developed in the last subsection, all parameters of the fuzzy rules can be estimated given a set of training samples. However, sufficient attention should be paid to the interpretability of such neurofuzzy systems during the data-driven learning. Problems that appear most often in such neurofuzzy systems are:

- Two neighboring fuzzy subsets in a fuzzy partition have no overlap and thus the fuzzy partition is incomplete.
- The membership functions of two fuzzy subsets are so similar that the distinguishability of the fuzzy partition is lost.
- The membership functions lose the prescribed physical meanings. For example, the center value of fuzzy set "small" is larger than that of "big", see Fig. 5.8.

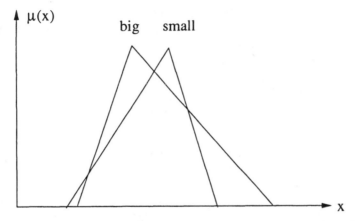

Fig. 5.8. The center of subset "small" is larger than that of "big".

- The fuzzy subset is too fuzzy or too crisp. This is caused by a too small or too large width of the fuzzy set.

To avoid these problems, constraints on the parameters of fuzzy membership functions should be introduced. At first, the following constraint on the centers of the membership functions are used:

$$a_j^1 < a_j^2 < \cdots < a_j^L, \tag{5.29}$$

where a_j^i is the center of the fuzzy subset $A_j^i(x_j), i = 1, 2, ..., L, 1 \leq L \leq N$ is the number of fuzzy sets in the fuzzy partition of input variable j. The order of the centers should be remain unchanged during learning.

One important point in maintaining good interpretability of TSK fuzzy systems is that the number of fuzzy subsets of an variable is usually much smaller than that of fuzzy rules. In other words, L should be smaller than N, especially when the number of fuzzy rules (N) is large. Heuristically, L should not be larger than 10. Therefore, the notation of the fuzzy subsets in the TSK fuzzy rule system in equation (5.2.1) is somewhat misleading because several of fuzzy subsets should have the same parameters.

One method to avoid an overly large number of fuzzy subsets in a fuzzy partition is to determine the structure of the neurofuzzy system beforehand.

The most popular structure is the grid structure. Unfortunately, the grid structure has a serious weakness. As the number of inputs in the fuzzy model increases, the needed number of fuzzy rules increases exponentially. An alternative to the grid structure is to identify the structure as proposed in [219].

In the following, an efficient approach to parameter estimation as well as structure identification of fuzzy system for neurofuzzy systems will be briefly described. It consists of the following steps:

1. Define an initial fuzzy partition for each variable.
2. Adjust parameters of the membership functions using the learning algorithm presented previously.
3. After the learning algorithm converges:
   - Sum up the firing strength (w^i) of each fuzzy rule over all training data:

   $$w^i = \sum_{k=1}^{M} w^i(k), \tag{5.30}$$

   where M is the number of training samples.
   - Normalize the firing strength:

   $$\bar{w}^i = w^i / \sum_{i=1}^{N} w^i, \tag{5.31}$$

   where N is the number of fuzzy rules.
   - Delete the fuzzy rule if

   $$\bar{w} < w^0, \tag{5.32}$$

   where w^0 is the prescribed minimal firing strength.
4. Check the similarity of all fuzzy subsets for each variable. The following fuzzy similarity measure [50] can be used for this purpose:

$$S(A_j^i(x_j), A_j^k(x_j)) = \frac{M(A_j^i(x_j) \cap A_j^k(x_j))}{M(A_j^i(x_j)) + M(A_j^k(x_j)) - M(A_j^i(x_j) \cap A_j^k(x_j))},$$
$$i \neq k, \tag{5.33}$$

where $M(\cdot)$ is called the size of a fuzzy set. Refer to Chapter 1 for more detailed discussions. Recall that if the similarity of two fuzzy sets is zero, then the two sets have no overlap. If the similarity between two fuzzy subsets is 1, then the two sets are the same. To reduce the number of fuzzy sets in a fuzzy partition, all fuzzy sets whose similarity is larger than a given threshold in a fuzzy partition are merged, which in fact also changes the fuzzy rule structure.

After the deletion of the inactive fuzzy rules and the merging of similar fuzzy subsets, training of the parameters is re-started. This process is repeated until a satisfying approximation performance as well as good interpretability is achieved.

5.3 Modeling and Control Using the Neurofuzzy System

5.3.1 Short-term Precipitation Prediction

Based on the historical data recorded during the years of 1952 and 1977 by Tianjin meteorological observatory [141], the western Pacific ocean temperature anomaly (x_1) from January the year before to February of the next year and the Eurasia 500 mb height anomaly (x_2) have been selected as two input variables of the neurofuzzy model for the prediction of precipitation in Tianjin. In this model, a standard grid rule structure is adopted. Each input variable is divided into 6 fuzzy subsets, namely, negative large (NL), negative medium (NM), negative small (NS), positive small (PS), positive medium (PM) and positive large (PL). The membership function of the fuzzy sets are as follows, which are shown in Fig. 5.9.

$$\mu_{NL}(z) = \begin{cases} 1, & z < -1, \\ \exp(-(z+1)^2/0.3), & -1 < z \le 0 \end{cases}.$$
$$\mu_{NM}(z) = \exp(-(z+0.5)^2/0.125), z < 0,$$
$$\mu_{NS}(z) = \exp(-z^2/0.125), z \le 0,$$
$$\mu_{PS}(z) = \mu_{NS},$$
$$\mu_{PM}(z) = \mu_{NM}(-z),$$
$$\mu_{PL}(z) = \mu_{NL}(-z). \tag{5.34}$$

In the training of the neurofuzzy model, the center values of PS and NS are fixed, and their widths are always equal. However, the symmetry between NL and PL, NM and PM will not be preserved.

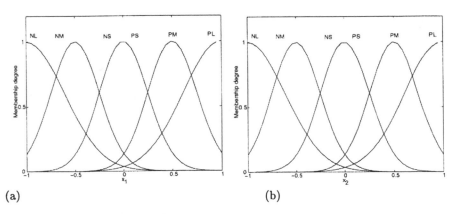

(a)   (b)

Fig. 5.9. Initial fuzzy membership functions. a) x_1, b) x_2.

For the sake of simplicity, the first order TSK fuzzy rules are adopted. According to the given fuzzy partition and the grid rule structure, 36 fuzzy rules can be generated:

R^1: If x_1 is NL and x_2 is NL, then $y^1 = P_0^1 + P_1^1 x_1 + P_2^1 x_2$;
R^2: If x_1 is NL and x_2 is NM, then $y^2 = P_0^2 + P_1^2 x_1 + P_2^2 x_2$;
$$: :$$
R^{36}: If x_1 is PL and x_2 is PL, then $y^{36} = P_0^{36} + P_1^{36} x_1 + P_2^{36} x_2$.

There are 26 data samples available for training from the year of 1952 to 1977. The neurofuzzy system is trained for 1000 iterations until the learning algorithm converges. Then the model is used to predict the rainfall from year of 1978 to 1981. Both the approximation and the prediction results are shown in Fig. 5.10. The largest discrepancy between the output of the neurofuzzy model and the desired output is in the year of 1977, which saw a high record of precipitation. The membership functions after learning are shown in Fig. 5.11.

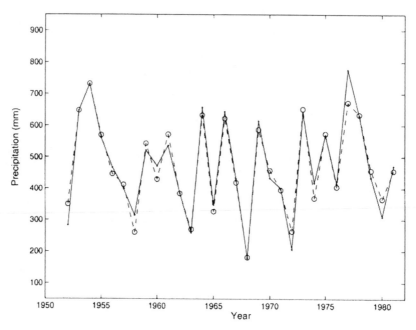

Fig. 5.10. Precipitation prediction. Solid line: Measured, dashed line: predicted.

5.3.2 Dynamic Robot Control

Dynamic control of robot manipulators has been widely investigated in the field of system control. Dynamic control of robot manipulators is a challenging task due to the fact that robot dynamics is highly nonlinear, strongly coupled, and full of uncertainties, including nonlinear friction, joint deflection, backlash and the unknown payload.

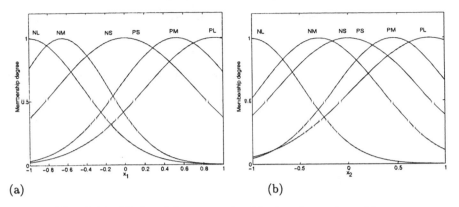

Fig. 5.11. Updated fuzzy membership functions. a) x_1, b) x_2.

Consider an N-degree-of-freedom rigid robot manipulator, whose dynamics is described by

$$\tau = H(q)\ddot{q} + M(q,\dot{q}) + G(q), \tag{5.35}$$

where τ is the $N \times 1$ torque vector, $H(q)$ is the $N \times N$ inertia matrix, $M(q,\dot{q})$ is the N-dimensional coriolis and centrifugal force vector, $G(q)$ is the N-dimensional gravity vector, and q, $\dot{q}$ and $\ddot{q}$ are the N-dimensional angular, velocity, and acceleration. Let $x_i = q_i$, $x_{N+i} = \dot{q}_i, i = 1, 2, ..., N$, then the robot dynamics can be rewritten by:

$$\dot{X} = A(X) + B(X)U, \tag{5.36}$$
$$Y = C(X), \tag{5.37}$$

where $X = [X_1 X_2]^T = [x_1, ..., x_N, x_{N+1}, ..., x_{2N}]^T$, $U = [\tau_1, ..., \tau_N]^T$, $C(X) = X_1$, and

$$A(X) = \begin{bmatrix} X_2 \\ -H^{-1}(M+G) \end{bmatrix}, \tag{5.38}$$

$$B(X) = \begin{bmatrix} 0 \\ -H^{-1} \end{bmatrix}. \tag{5.39}$$

Before the neurofuzzy model can be applied to the control of the robot system governed by equations (5.36) and (5.37), the robot dynamics must be decoupled beforehand. Define the following nonlinear operator [66]:

$$N_A^j C_i(X) = \left[\frac{\partial}{\partial X} N_A^{j-1} C_i(X) \right] A(X), j = 1, 2, ..., N-1, \tag{5.40}$$

$$N_A^0 C_i(X) = C_i(X), \tag{5.41}$$

where $C_i(X)$ is the i-th row of $C(X)$. Define the relative degree of the system:

$$d_i = \min_j \left\{ \left[\frac{\partial}{\partial X} N_A^{j-1} C_i(X) \right] B(X) \neq 0 \right\}, j = 1, 2, ..., N. \tag{5.42}$$

Then the following control input is able to decouple the robot dynamics:

$$U = F(X) + G(X)V, \tag{5.43}$$

where V is the new control vector of the decoupled linear system, and

$$F(X) = -(D^*)^{-1}(X)(F_1^*(X) + F_2^*(X)), \tag{5.44}$$
$$G(X) = -(D^*)^{-1}(X)\Lambda, \tag{5.45}$$
$$D_i^*(X) = \frac{\partial}{\partial X}[N_A^{d_i}C_i(X)]B(X), \tag{5.46}$$
$$F_{1i}(X) = N_A^{d_i}C_i(X), \tag{5.47}$$
$$F_{2i}(X) = \sum_{k=1}^{d_i-1} a_{k,i} N_A^k C_i(X), \tag{5.48}$$
$$\Lambda = \text{diag}[\lambda_1, ..., \lambda_N], \tag{5.49}$$

where D_i^*, $F_{1i}(X)$, and $F_{2i}(X)$ are the i-th row of matrix $D^*(X)$, $F_1^*(X)$, and $F_2^*(X)$, respectively, and $a_{k,i}$ are certain constants to be determined. For the robot system given in equation (5.36), since

$$\frac{\partial}{\partial X}[N_A^0 C_i(X)]B(X) = 0, \tag{5.50}$$

and

$$\frac{\partial}{\partial X}[N_A^1 C_i(X)]B(X) \neq 0, \tag{5.51}$$

thus, the relative degree of the system is

$$d_i = 2, i = 1, 2, ..., N. \tag{5.52}$$

In this way, the robot system is transformed into the following decoupled linear system:

$$\begin{bmatrix} \dot{x}_i \\ \vdots \\ \dot{x}_N \\ \dot{x}_{N+1} \\ \vdots \\ \dot{x}_{2N} \end{bmatrix} = \begin{bmatrix} x_{N+1} \\ \vdots \\ x_{2N} \\ -a_{0,1}x_1 - a_{1,1}x_N \\ \vdots \\ -a_{0,N}x_N - a_{1,N}x_{2N} \end{bmatrix} + \begin{bmatrix} 0 \\ \lambda_1 \\ & \ddots \\ & & \lambda_N \end{bmatrix} \begin{bmatrix} v_1 \\ \vdots \\ v_N \end{bmatrix}. \tag{5.53}$$

It is straightforward that the subsystem for each link of the manipulator is a time-invariant two-input single-output linear system. The parameters $a_{k,i}$ and λ_i should be chosen in such a way that each linear subsystem is stable. A widely used decoupled model looks like

$$\gamma \ddot{q}_i + s\dot{q}_i = v_i. \ (\gamma, s > 0, i = 1, 2, ..., N) \tag{5.54}$$

However, the subsystems are linear time-invariant only if the parameters of the robot dynamics are exactly known and no other nonlinear uncertainties exist. Unfortunately, this is not true in real applications. As a result, a conventional PID controller will most probably fail to achieve a high performance on the decoupled system. Therefore, an adaptive fuzzy system will be developed to deal with the uncertainties in robot dynamics.

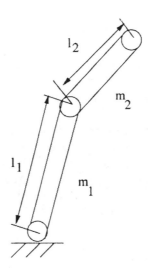

Fig. 5.12. A two-degree planar robot.

As an illustrative example, the following two-degree-of-freedom rigid manipulator is considered in the simulation, as shown in Fig. 5.12.

$$\begin{bmatrix} \tau_1 \\ \tau_2 \end{bmatrix} = \begin{bmatrix} H_{11} & H_{12} \\ H_{21} & H_{22} \end{bmatrix} \begin{bmatrix} \ddot{q}_1 \\ \ddot{q}_2 \end{bmatrix} + \begin{bmatrix} M_{11} & M_{12} \\ M_{21} & M_{22} \end{bmatrix} \begin{bmatrix} \dot{q}_1 \\ \dot{q}_2 \end{bmatrix} + \begin{bmatrix} G_1 \\ G_2 \end{bmatrix}, \tag{5.55}$$

where

$$H_{11} = (m_1 + m_2)l_1^2 + m_2 l_2^2 + 2m_2 l_1 l_2 \cos(q_2), \tag{5.56}$$

$$H_{12} = H_{21} = m_2 l_2^2 + m_2 l_1 l_2 \cos(q_2), \tag{5.57}$$

$$H_{22} = m_2 l_2^2, \tag{5.58}$$

$$M_{11} = -2m_2 l_1 l_2 \sin(q_2)\dot{q}_2, \tag{5.59}$$

$$M_{12} = -m_2 l_1 l_2 \sin(q_2)\dot{q}_2, \tag{5.60}$$

$$M_{21} = m_2 l_1 l_2 \sin(q_2)\dot{q}_1, \tag{5.61}$$

$$M_{22} = 0, \tag{5.62}$$

$$G_1 = [(m_1 + m_2)l_1 \sin(q_1) + m_2 l_2 \sin(q_1 + q_2)]g, \tag{5.63}$$

$$G_2 = m_2 l_2 \sin(q_1 + q_2)g, \tag{5.64}$$

where m_1 and m_2 are the mass of the two links, l_1 and l_2 are the link lengths, and g is the gravity.

The decoupled model for each link of the manipulator is set as follows:

$$0.1\ddot{q}_i + \dot{q}_i = v_i \ (i = 1, 2). \tag{5.65}$$

Choose $x_1^i = \dot{q}_i$, $x_2^i = q_i^d - q_i$ as the inputs of the fuzzy model, and $y^i = v_i$ as the output of the model.

When a neurofuzzy system is employed for system control, an additional problem need to be solved. In fuzzy modeling, the desired output of the fuzzy model is known for a given input from the training data set. Therefore, the gradient-based learning algorithm developed in the previous section can directly be applied. However, in fuzzy control, the desired output of the fuzzy model is unknown, although the desired output of the controlled plant is usually given. To address this problem, two approaches may be adopted. For example, the reinforcement learning can be used [215]. A more feasible method is to estimate the derivative of the output with respect to the input of the controlled plant. In this example, the desired value for v_i is unknown, although the desired output of the system (q_i) is known. Consider the following cost function of the fuzzy controller:

$$E = \frac{1}{2}(q_i(t) - q^d(t))^2. \tag{5.66}$$

Let θ represent any parameter of the neurofuzzy model,

$$\frac{\partial E}{\partial \theta} = (q_i - q_i^d)\frac{\partial q_i}{\partial v_i}\frac{\partial v_i}{\partial \theta}, \tag{5.67}$$

where $\partial v_i/\partial \theta$ can be derived from the neurofuzzy model, $\partial q_i/\partial v_i$ needs to be estimated from the dynamics of the controlled plant. If the analytical model of the plant is known, it is easy to estimate the derivative. On the other hand, it is also feasible to estimate the partial derivative using the following discrete form:

$$\frac{\partial q_i}{\partial v_i} \approx \frac{q_i(t) - q_i(t-1)}{v_i(t) - v_i(t-1)}. \tag{5.68}$$

In this way, the neurofuzzy model can be employed in system control without any problem.

Suppose the desired trajectory for link 1 and 2 is given as follows:

$$q_1^d(t) = \exp(0.5t) \ (rad), \quad q_1(0) = 0.8 \ (rad), \tag{5.69}$$
$$q_2^d(t) = 0.5 + \exp(0.4t) \ (rad), \quad q_2(0) = 1.4 \ (rad). \tag{5.70}$$

To simulate the model uncertainties in real applications, the following nonlinear friction has been taken into account in the simulation, which is supposed to not have been modeled in the robot dynamics:

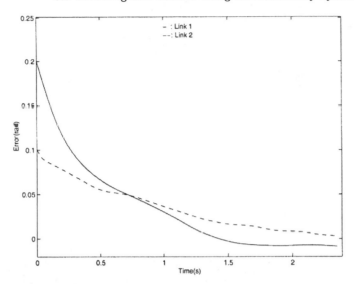

Fig. 5.13. Tracking errors after 10 learning cycles. Solid line for link 1; and dashed line for link 2.

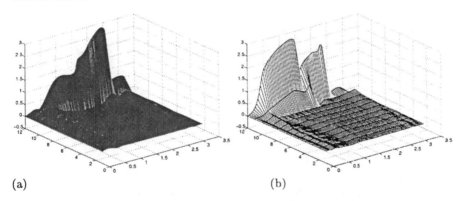

(a)   (b)

Fig. 5.14. Training process. a) Link 1, b) Link 2.

$$f_i = f_{q_i}(\dot{q}_i, \tau_i) + f_{v_i}(\dot{q}_i), i = 1, 2, \tag{5.71}$$

where f_{q_i} and f_{v_i} are the Columbus and viscous friction, respectively, which can be expressed by

$$f_{q_i}(\dot{q}_i, \tau_i) = \begin{cases} k_i \mathrm{sgn}(\dot{q}_i), & |\dot{q}_i| > 0, \\ k_i \mathrm{sgn}(\tau_i), & \dot{q}_i = 0, |\tau_i| > k_i, \\ \tau_i, & \dot{q}_i = 0, |\tau_i| < k_i, \end{cases} \tag{5.72}$$

$$f_{v_i}(\dot{q}_i) = C_i \dot{q}_i, \tag{5.73}$$

where k_i and C_i are known constants.

In the simulation, all input variables are normalized between $[-1, 1]$, which is partitioned into six fuzzy subsets $\{NL, NM, NS, PS, PM, PL\}$. The tracking errors of link 1 and link 2 are shown in Fig. 5.13 after 10 learning iterations. The training process of link 1 and link 2 is shown in Fig. 5.14, where the x-coordinate is the time, the y-coordinate is the training iterations and the z-coordinate is the training error. The membership functions after learning are shown in Fig. 5.15 and Fig. 5.16.

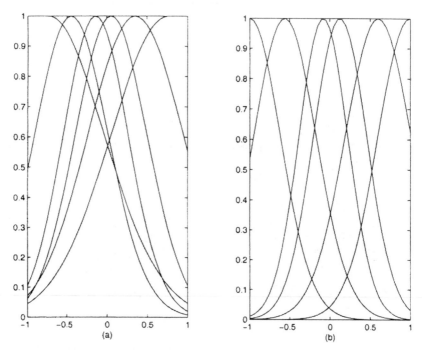

Fig. 5.15. Membership functions for link 1. a) x_1, b) x_2.

5.4 Neurofuzzy Control of Nonlinear Systems

Control of nonlinear systems is a challenging topic in the field of system control. As it is well known, a basic approach to the control of nonlinear systems is to linearize the given system and then employ a control strategy developed for linear systems. Generally speaking, there are two main approaches to system linearization, namely, approximate linearization and exact linearization. The most widely used approximate linearization method is to linearize the original system around its operating point using Taylor expansion. After that, either a conventional linear controller , for example, a PID controller,

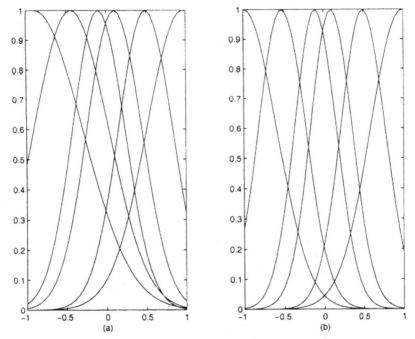

Fig. 5.16. Membership functions for link 2. a) x_1, b) x_2.

is applied [256] or a self-tuning controller is constructed by virtue of on-line identification [155]. Another approximate linearization method is to replace the original system with a group of piecewise linear systems [206]. However, for a class of substantially nonlinear systems, the approximate linearization methods are subject to large approximation errors and therefore satisfying control performance can hardly be obtained.

The differential geometry theory that was first developed in the 1970s provided an alternative way for nonlinear system control. By taking advantage of the diffeomorphism transformation and the nonlinear state feedback, a class of nonlinear systems can be exactly linearized [106]. Unfortunately, this class of methods require an analytical mathematical model of the original system and are very sensitive to the model quality.

In this section, a generic framework for nonlinear systems control using neurofuzzy models is presented. With the help of the neurofuzzy model introduced previously, a set of fuzzy linear subsystems to model a given nonlinear system. A linear optimal controller is then designed for each subsystems. The control action of each subsystem is then synthesized using fuzzy composition to control the nonlinear system . In this way, the problem of nonlinear control is reduced to linear system control. The stability of the whole fuzzy system is discussed under two different cases by means of Lyapunov second method.

5.4.1 Fuzzy Linearization

Consider the following nonlinear system:

$$x_k = f(x_{k-1}, u_{k-1}), \tag{5.74}$$

where x_k, $x_{k-1} \in R^{n \times 1}$ are the system states and $u_{k-1} \in R^{m \times 1}$ is the control vector of the system. If system (5.74) is expanded at $x_{k-1} = x_{k-1}^*$ and $u_{k-1} = u_{k-1}^*$ with the Tailor series, then we have

$$\Delta x_k = A_{k-1} \Delta x_{k-1} + B_{k-1} \Delta u_{k-1}, \tag{5.75}$$

where,

$$A_{k-1} = \frac{\partial f}{\partial x_{k-1}}|_*, \ B_{k-1} = \frac{\partial f}{\partial u_{k-1}}|_*, \tag{5.76}$$

where $A_{k-1} \in R^{n \times n}$, $B_{k-1} \in^{n \times m}$ are time-varying. If system (5.74) is sufficiently smooth, (A_{k-1}, B_{k-1}) can be treated as time-invariant within a period of time and therefore the following piecewise linear time-invariant (LTI) models can be obtained (refer to Fig. 5.17 for one-dimensional case):

$$x_k = A_1 x_{k-1} + B_1 u_{k-1}, \text{ if } a_1 \leq x_{k-1} \leq a_2$$
$$x_k = A_2 x_{k-1} + B_2 u_{k-1}, \text{ if } a_2 \leq x_{k-1} \leq a_3$$
$$\vdots$$
$$x_k = A_p x_{k-1} + B_p u_{k-1}, \text{ if } a_p \leq a_{k-1} \leq a_p. \tag{5.77}$$

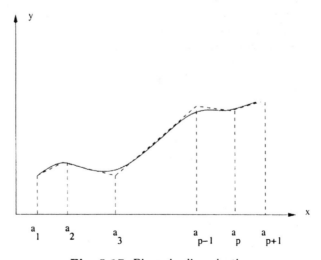

Fig. 5.17. Piecewise linearization.

In this case, (A_j, B_j) $(j = 1, 2, ..., p)$ are LTI systems. According to the piecewise linear models in equation (5.77), a linear controller can be designed for each subsystem to control the nonlinear model. However, we can see from Fig. 5.17 that the boundaries for each linear model $[a_1, a_2], ..., [a_p, a_{p+1}]$ are crisp, which unavoidably causes large approximation errors near the boundary. As a matter of fact, it can be seen that the original system does not have these boundaries inherently. If it has, the boundaries are fuzzy but not crisp. For this reason, fuzzy boundaries can be defined as in Fig. 5.18, which is able to significantly reduce the approximating error around the boundary. Therefore, the nonlinear system (5.74) is fuzzily but much more exactly linearized compared to the piecewise linearization method. Note that the definite ranges $[a_1, a_2], [a_2, a_3], ..., [a_p, a_{p+1}]$ have become a fuzzy partition consisting of a number of fuzzy subsets $F_1, F_2, F_3, ..., F_p$ defined by a membership function.

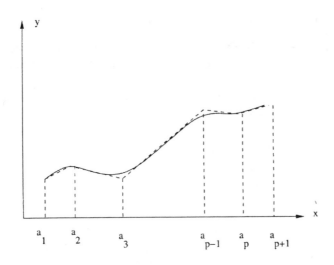

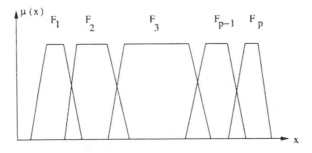

Fig. 5.18. Fuzzy linearization.

To facilitate further discussions, it is assumed that the nonlinear system has the following properties:

1. Control u_{k-1} is bounded, i.e., there exists a positive constant vector u_0, such that

$$\| u_{k-1} \| \leq u_0, 1 \leq k < \infty. \tag{5.78}$$

2. All system states are measurable.
3. The nonlinear system is bounded, i.e., there exist constant vectors M and N such that $\forall \| x_0 \| \leq M, \| x_k \| \leq N$.
4. On the compact set U, the nonlinear function $f(\cdot)$ is a real, continuous function.

On the basis of the above discussion, the nonlinear system (5.74) can the described using the following P fuzzy linear models:

$$R^j: \text{If } x_{k-1} \text{ is } F^j, \text{ then } x_k^j = A_{k-1}^j x_{k-1} + B_{k-1}^j u_{k-1},$$

where, R^j denotes the j-th fuzzy rule, (A_{k-1}^j, B_{k-1}^j) represents the j-th fuzzy linear subsystem. If system (5.74) has no uncertainties, then the subsystems (A_{k-1}^j, B_{k-1}^j) are time-invariant. Thus, the control of nonlinear system (5.74) is reduced to the control of fuzzy subsystems. According to the TSK inference method, the subsystems can be combined into one single system as follows:

$$x_k = \frac{\sum_{j=1}^P (w^j x_{k-1}^j)}{\sum_{j=1}^P w^j},$$

$$= \frac{\sum_{j=1}^P (w^j A_{k-1}^j)}{\sum_{j=1}^P w^j} x_{k-1} + \frac{\sum_{j=1}^P (w^j B_{k-1}^j)}{\sum_{j=1}^P w^j} u_{k-1}, \tag{5.79}$$

where w^j is the truth value (firing strength) of the j-th rule:

$$w^j = \mu_{F^j}(x_{k-1}), \tag{5.80}$$

which satisfies $w^j \geq 0$ and $\sum_{j=1}^P w^j > 0$.

For the nonlinear system (5.74) and the fuzzy system (5.79), we have the following observations without proof.

- If the nonlinear system is a deterministic system without uncertainties, all fuzzy linear subsystems and their membership functions are time-invariant. Once the parameters of each linear subsystem is estimated, a linear controller can be designed.
- The fuzzy linear subsystems have a better approximation accuracy than the piecewise linear systems in equation (5.77). Therefore, a controller designed based on the fuzzy subsystems has better performance than the one based on the piecewise linear systems.

- If the original system is a stochastic system, the linear subsystems will be time varying. It is expected that the parameters of the fuzzy linear subsystems change slower than those in the linear model obtained using the Tailor expansion. Under certain conditions, the fuzzy linear subsystems can be treated as slowly time-varying systems.

Although the previous observations cannot be proved strictly, the following theorem [235] on the uniform approximation capability of TSK fuzzy systems can be seen as an indirect proof.

Theorem For a class of TSK fuzzy systems as expressed in equation (5.2.1), if

1. all fuzzy membership functions are Gaussian;
2. the t-norm is the multiplication,

then for any continuous function on the compact set U, there exists a fuzzy system $f(x)$ such that the following relation holds:

$$\sup_{x \in U} |g(x) - f(x)| < \epsilon, \qquad (5.81)$$

where ϵ is an arbitrarily small positive number. Refer to [235] for a detailed proof of the theorem.

From the above theorem, it is obvious that the fuzzy linear subsystems are able to approximate the original nonlinear system to an arbitrary accuracy, which guarantees the convergence of the estimation algorithm for the fuzzy linear subsystems.

5.4.2 Neurofuzzy Identification of the Subsystems

It can be seen that the fuzzy linear subsystems in equation (5.79) is an extension of the TSK model in equation (5.2.1), therefore, it is feasible to estimate the parameters of the linear subsystems using the neurofuzzy model described in Section 5.2.1.

Let: $x^T(k) = (x_1, x_2, ..., x_n)^T$, $u^T(k) = (u_1, u_2, ..., u_m)^T$, and

$$A^j_{k-1} = \begin{bmatrix} a^j_{11} & a_{12} & \cdots & a^j_{1n} \\ a^j_{21} & a_{22} & \cdots & a^j_{2n} \\ \vdots & \vdots & \ddots & \vdots \\ a^j_{n1} & a_{n2} & \cdots & a^j_{nn} \end{bmatrix} \qquad (5.82)$$

$$B^j_{k-1} = \begin{bmatrix} b^j_{11} & b_{12} & \cdots & b^j_{1n} \\ b^j_{21} & b_{22} & \cdots & b^j_{2n} \\ \vdots & \vdots & \ddots & \vdots \\ b^j_{n1} & b_{n2} & \cdots & b^j_{nn} \end{bmatrix} . \qquad (5.83)$$

For the sake of convenience, we change the notation of the system as follows:

$$c_{1,1}^j = a_{11}^j, \cdots, c_{1,n}^j = a_{1n}^j, \ c_{1,n+1}^j = b_{11}^j, \cdots, c_{1,n+m}^j = b_{1m}^j, \qquad (5.84)$$

$$\vdots \qquad\qquad\qquad \vdots \qquad\qquad\qquad (5.85)$$

$$c_{n,1}^j = a_{n1}^j, \cdots, c_{n,n}^j = a_{nn}^j, \ c_{n,n+1}^j = b_{n1}^j, \cdots, c_{n,n+m}^j = b_{nm}^j. \qquad (5.86)$$

Thus, the fuzzy system can be expressed by

$$R^j: \text{If } x_1^j(k-1) \text{ is } F_1^j, \cdots, x_n^j(k-1) \text{ is } F_n^j \text{ then}$$
$$x_1^j(k) =$$
$$c_{1,1}^j x_1(k-1) + \cdots + c_{1,n}^j x_n(k-1) + c_{1,n+1}^j u_1(k-1) + \cdots + c_{1,n+m}^j u_m(k-1);$$
$$\cdots$$
$$x_n^j(k) =$$
$$c_{n,1}^j x_n(k-1) + \cdots + c_{n,n}^j x_n(k-1) + c_{n,n+1}^j u_n(k-1) + \cdots + c_{n,n+m}^j u_m(k-1),$$

which is in the form of first order TSK fuzzy model. Define the cost function:

$$E = \frac{1}{2}(x_i(k) - x_i^d(k))^2, i = 1, 2, ..., n, \qquad (5.87)$$

and use the gradient method:

$$\frac{\partial E}{\partial c_{i,r}^j} = \frac{\partial E_i}{\partial x_i(k)} \frac{\partial x_i(k)}{\partial c_{i,r}^k}$$

$$= [x_i(k) - x_i^d(k)]w_i^j z_r / \sum_{j=1}^p w_i^j, \qquad (5.88)$$

where $i = 1, 2, ..., n, r = 1, 2, ..., n + m$, and

$$z_1 = x_1(k-1) \qquad\qquad (5.89)$$
$$z_2 = x_2(K-1) \qquad\qquad (5.90)$$
$$\vdots \qquad\qquad (5.91)$$
$$z_n = x_n(k-1) \qquad\qquad (5.92)$$
$$z_{n+1} = u_1(k-1) \qquad\qquad (5.93)$$
$$\vdots \qquad\qquad (5.94)$$
$$z_{n+m} = u_m(k-1). \qquad\qquad (5.95)$$

Therefore, the learning algorithm for the model parameters is:

$$\Delta c_{i,r}^j = -\lambda[x_i(k) - x_i^d(k)]w_i^j z_r / \sum_{j=1}^p w_i^j. \qquad (5.96)$$

The learning algorithm for the membership functions can be derived similarly. Recall that multiplication is used in this case.

5.4.3 Design of Controller

For each linear subsystem, if (A^j_{k-1}, B^j_{k-1}) is controllable, then an optimal feedback controller can be designed:

R^j: If $x_1(k-1)$ is F^j_1, and ..., and $x_n(k-1)$ is F^j_n, then

$$x^j_k = A^j_{k-1}x_{k-1} + B^j_{k-1}u_{k-1}, \tag{5.97}$$

$$u^j_{k-1} = -L^j_{k-1}x_{k-1}, \tag{5.98}$$

where L^j_{k-1} is the optimal feedback gain. Thus, the whole system becomes:

$$x_k = [\sum_{j=1}^{p}(w^j A^j_{k-1})x_{k-1} + \sum_{j=1}^{p}(w^j B^j_{j-1})u_{k-1}]/\sum_{j=1}^{p}w^j, \tag{5.99}$$

$$u_{k-1} = -\sum_{j=1}^{p}(w^j L^j_{k-1})x_{k-1}/\sum_{j=1}^{p}w^j. \tag{5.100}$$

Combining equations (5.99) and (5.100), we have

$$x_k = \frac{\sum_{i=1}^{p}\sum_{j=1}^{p}w^i w^j[A^j_{k-1} - B^j_{k-1}L^i_{k-1}]x_{k-1}}{\sum_{i=1}^{p}\sum_{j=1}^{p}w^i w^j}, \tag{5.101}$$

$$= \frac{1}{R}[\sum_{j=1}^{p}w^j w^j G^{jj}x_{k-1} + 2\sum_{j<i}w^j w^i\frac{G^{ij} + G^{ij}}{2}x_{k-1}], \tag{5.102}$$

where,

$$G^{ji} = A^j - B^j L^i_{k-1}, \quad R = \sum_{i=1}^{p}\sum_{j=1}^{p}w^i w^j. \tag{5.103}$$

Equation (5.103) can be rewritten as:

$$x_k = \frac{\sum_{j=1}^{p(p+1)/2}(v^j M^j x_{k-1})}{\sum_{j=1}^{p(p+1)/2}v^j}, \tag{5.104}$$

where,

if $j = i$, $\quad M^{\sum_{t=1}^{i}(t-1)+j} = G^{ji}, \qquad v^{\sum_{t=1}^{i}(t-1)+j} = w^j w^i$ (5.105)

if $j < i$, $M^{\sum_{t=1}^{i}(t-1)+j} = (G^{ji} + G^{ij})/2, v^{\sum_{t=1}^{i}(t-1)+j} = 2w^j w^i$ (5.106)

The above notation of the combined fuzzy system paves the way for the stability analysis of the whole fuzzy system.

5.4.4 Stability Analysis

For the sake of simplicity, equation (5.104) can be rewritten as:

$$x_k = \frac{\sum_{j=1}^{L}(v^j M_{k-1}^j)x_{k-1}}{\sum_{j=1}^{L} v^j}, \tag{5.107}$$

where $L = p(p+1)/2$.

Lemma [220] For any positive definite matrix P, which satisfies

$$A^T A - P < 0, \text{ and } B^T P B - P < 0, \tag{5.108}$$

then

$$A^T PB + B^T PA - 2P < 0, \tag{5.109}$$

where $A, B, P \in R^{n \times n}$. For the proof of the lemma, see [220].

Theorem [220] Fuzzy system (5.107) is globally asymptotically stable if there exist a common positive definite matrix P for each LTI subsystem M^j such that

$$(M^j)^T PM^j - P < 0, j = 1, 2, ..., L. \tag{5.110}$$

See [220] for the proof of the theorem.

As previously discussed, the linear subsystems are slowly time-varying, if the original nonlinear system (5.74) is not deterministic. Therefore, the stability properties of fuzzy system (5.104) is very important when the linear subsystems are slowly time-varying.

Theorem Fuzzy system (5.107) is exponentially stable at $x_k = 0$, if

1. There exist two constants a and b and a symmetric matrix P_k for all M^j such that:

$$0 < aI \le P_k \le bI \infty \text{holds for all } k; \tag{5.111}$$

2. There exists a common matrix H_k such that

$$(M_k^j)^T P_{k+1} M_k^j - P_k = -H_k H_k^T; \tag{5.112}$$

3. There exists a constant C and a function $S_n(k)$, for a certain n and for all k such that:

$$S_n(k) = \sum_{i=0}^{n-1} \Phi^t(k+i,k)H(k+i)H^T(k+i)\Phi(k+i,k) \ge CI > 0, \tag{5.113}$$

where $\Phi(\cdot)$ is the transfer matrix of system M_k.

Proof Select the following Lyapunov function:

$$V(x_k, k) = x_k^T P_k x_k, \tag{5.114}$$

where P_k satisfies the conditions in (5.108). Thus,

1. If $x_k = 0$, then $V(0, k) \equiv 0$;
2. If $x_k \neq 0$, then $0 < a \parallel x_k \parallel^2 \leq V(x_k, k) \leq b \parallel x_k \parallel^2$;
3. If $x \to \infty$, them $V(x_k, k) \to \infty$.

Then,

$$
\begin{aligned}
\Delta V &= x_{k+1}^T P_{k+1} x_{k+1} - X_k^T P_k X_k \\
&= \left[\frac{\sum_{j=1}^L (v^j M_k^j)}{\sum_{j=1}^L v^j} x_k \right]^T P_{k+1} \frac{\sum_{j=1}^L (v^j M_k^j)}{\sum_{j=1}^L v^j} x_k - x_k^T P_k x_k \\
&= x_k^T \left[\frac{\sum_{j=1}^L v^j (M_k^j)^T}{\sum_{j=1}^L v^j} P_{k+1} \frac{\sum_{j=1}^L (v^j M_k^j)}{\sum_{j=1}^L v^j} - P_k \right] x_k \\
&= \frac{1}{[\sum_{j=1}^L v^j]^2} \left\{ \sum_{j=1}^L (v^j)^2 x_k^T [(M_k^j)^T P_{k+1} M_k^j M_k^j - p_k] x_k \right\} \\
&\quad + \sum_{i=1}^L \sum_{j=1}^L \left\{ v^i v^j x_k^T \left[(M_k^i)^T P_{k+1} M_k^j + (M_k^j)^T P_{k+1} M_k^i - 2P_k \right] x_k \right\}.
\end{aligned}
$$

$$(5.115)$$

According to the condition 2, we have

$$
\begin{aligned}
\Delta V &= \frac{1}{\left[\sum_{j=1}^L v^j \right]^2} \left\{ -\sum_{j=1}^L (v^j)^2 x_k^T H_k H_k^T x_k - 2 \sum_{i=1}^L \sum_{j=i+1}^L [v^i v^j x_k^T H_k H_k^T x_k] \right. \\
&\quad \left. - \sum_{i=1}^L \sum_{j=i+1}^L [v^i v^j x_k^T (M_k^i - M_k^j)^T P_k (M_k^i - M_k^j)] x_k \right\} \\
&= \left\{ -\sum_{j=1}^L (v^j)^2 x_k^T H_k H_k^T x_k \right. \\
&\quad \left. - \sum_{i=1}^L \sum_{j=i+1}^L v^i v^j x_k^T (M_k^i - M_k^j)^T P_k (M_k^i - M_k^j) x_k \right\} / \left[\sum_{j=1}^L v^j \right]^2 \\
&\leq 0,
\end{aligned}
$$

$$(5.116)$$

where $v^j \geq 0$, $\sum_{j=1}^L v^j > 0$. Obviously, $v(x_k, k)$ is decreasing. Furthermore,

$$\Delta V(x, k+1) = -x_{k+1}^T H_{k+1} H_{k+1}^T x_{k+1}$$

$$-\sum_{i=1}^{L} \sum_{j=i+1}^{L} \left\{ v^i v^j x_{k+1}^T (M_{k+1}^i - M_{k+1}^j)^T P_{k+1} \right.$$

$$\left. (M_{k+1}^i - M_{k+1}^j) x_{k+1} \right\} / (\sum_{j=1}^{L} v^j)^2$$

$$= -x_k^T \Phi^T(k+1, k) H_{k+1}^T x_k^T \Phi^T(k+1, k)$$

$$-\sum_{i=1}^{L} \sum_{j=i+1}^{L} \left\{ v^i v^j x_k^T \Phi^T(k+1, k)(M_{k+1}^i - M_{k+1}^j)^T P_{k+1} \right.$$

$$\left. (M_{k+1}^i - M_{k+1}^j) x_k^T \Phi^T(k+1, k) \right\} / (\sum_{j=1}^{L} v^j)^2. \qquad (5.117)$$

Therefore,

$$\Delta V(x, k+N-1) = x_{k+N-1}^T P_{k+N-1} - x_k^T P_k x_k$$

$$= \sum_{l=0}^{n-1} \Delta V(x, k+1)$$

$$= -\sum_{l=1}^{N-1} \left\{ v^i v^j x_k^T \Phi^T(k+1, k)(M_{k+1}^i - M_{k+1}^j)^T \right.$$

$$\left. P_{k+1}(M_{k+1}^i - M_{k+1}^j) x_k^T \Phi(k+1, k) \right\} / \sum_{j=1}^{L} (v^j)^2. $$

$$(5.118)$$

According to the condition 2 of the theorem, we have

$$\Delta V(x, k+N-1) \leq -x_k^T S_n(k) x_k$$

$$-\sum_{l=1}^{N-1} \left\{ v^i v^j x_k^T \Phi^T(k+1, k)(M_{k+1}^i - M_{k+1}^j)^T \right.$$

$$\left. P_{k+1}(M_{k+1}^i - M_{k+1}^j) x_k^T \Phi(k+1, k) \right\} / \sum_{j=1}^{L} (v^j)^2$$

$$\leq -x_k^T C x_k. \qquad (5.119)$$

Thus,

$$\frac{\Delta V(x, k-1+N)}{V(x, k-1)} \leq \frac{-C \parallel x_k \parallel^2}{-a \parallel x_k \parallel^2} = -\frac{C}{a} < 0, \qquad (5.120)$$

which means that $V(x, k)$ converges to zero exponentially. Therefore, fuzzy system (5.107) is exponentially stable.

5.5 Summary

After a short review of the main existing neurofuzzy models, this chapter presents a neurofuzzy model based on the pi-sigma hybrid neural network. This neurofuzzy model can be seen as a neural network implementation of a nonlinear TSK fuzzy model. Another feature of the model is that it is able to use minimum for fuzzy intersection. To this end, learning algorithms for training this neurofuzzy model have been proposed. To illustrate the functionality of the neurofuzzy model, it has been applied to modeling and control of complex systems. Finally, linearization of nonlinear systems with the help of the neurofuzzy model has been studied. Proof of stability of the linearized system is also given based on the Lyapunov method.

6. Evolutionary Design of Fuzzy Systems

6.1 Introduction

Neurofuzzy systems are adaptive fuzzy systems that are able to learn using learning methods developed in the field of neural networks. Besides, neurofuzzy systems are assumed to have all important features of fuzzy systems, i.e., the knowledge represented by a neurofuzzy system should be transparent to human users.

Although neurofuzzy systems are able to self-organize their structure in some cases, the structure of most neurofuzzy systems is defined beforehand. However, the structure of fuzzy systems is sometimes essential for the performance. Besides, most neurofuzzy systems use learning algorithms based on the gradient methods, which have several drawbacks. For example, they require the derivative information of the system, which causes serious difficulties for design of fuzzy controllers, where the mathematical model of the controlled plant is often unknown.

Evolutionary algorithms are heuristic global optimization methods imitating the main mechanisms of the biological evolution. The main merits of evolutionary algorithms are the following. First, they do not need to know the exact mathematical model of the system to be optimized. The only information needed is the performance of a given design. This is very desirable in the design of fuzzy controllers. Second, they do not require that the system to be optimized is derivable, which makes it possible to apply evolutionary algorithms to the optimization of non-derivable systems, such as the optimization the table-look-up fuzzy systems and most importantly, the optimization of the structure of fuzzy systems.

There are a huge number of papers that combine fuzzy logic with evolutionary algorithms. Main differences of the methods lie in the following aspects:

- Parameter/structure optimization. If we take a look at the structure of a fuzzy logic system, refer to Fig. 6.1, we can see that several elements in the fuzzy system can be optimized. For example, in the simplest case, evolutionary algorithms can be used to optimize the parameters of the membership functions for both input and output linguistic variables. In case TSK fuzzy rules are used, then model parameters instead of parameters of output membership functions are to be optimized. In this case, the number of fuzzy subsets in each fuzzy partition is fixed and only the shape of the membership functions are adjustable. To go a step further, not only the parameters, but also the structure (the partition) of the linguistic variables can also be optimized. That is to say, the number of fuzzy subsets in each partition is to be evolved. In the these fuzzy systems, the grid structure is assumed, which means that if there are n input variables and each variable has m_i fuzzy subsets, then there will be $N = \prod_{i=1}^{n} m_i$ fuzzy rules. Since the grid structure may cause "rule explosion" in high-dimensional systems, this structure can also be discarded and an optimal rule structure is to be searched using evolutionary algorithms. To realize the parameter as well as the structure optimization, a flexible genetic representation of the fuzzy system may be necessary, including variable chromosome lengths, messy genetic algorithms [91] and context dependent coding [156]. Alternatively, genetic programming can also be employed [228].

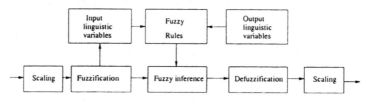

Fig. 6.1. A generic structure of fuzzy systems.

- Michigan/Pittsburg approach. In most approaches to evolutionary fuzzy systems design, the Pittsburg approach of evolution is used. In this approach, the entire fuzzy system is encoded in one chromosome and each individual in the population competes with each other during evolution. Finally, the best individual is the optimization result. In the Michigan approach, a chromosome encodes only one fuzzy rule and the whole population consists of the fuzzy rule base. In this case, the individuals in the population are believed to co-operate with each other to achieve a good performance. However, difficulties arise in fitness assignment. To alleviate this difficulty, an iterative approach has been proposed in [40]. In this method, a chromosome represents a fuzzy rule, and only one fuzzy rule is generated in one run. The optimization is run iteratively so that a number of fuzzy rules can be found which have a good collective performance. In [107], a subset of the population randomly selected to be evaluated and this fitness is assigned to each individuals in the subset. This process is repeated several times and thus the average fitness of an individual over different evaluations is assigned to the individual.
- Single/multiobjective. The most popular criterion for fuzzy systems design is the quadratic error function for modeling or control applications. Therefore, most evolutionary fuzzy systems have one objective. However, this is

insufficient in many cases. One of the most objective in addition to the error function is the interpretability condition. In [126], evolution strategies are used to generate fuzzy systems that are flexible, complete, consistent and compact. Conventional weighted sum method is used to combine different objectives into one single objective. Pareto-based approaches to multiobjective optimization of fuzzy systems have also been proposed [104, 112].

6.2 Evolutionary Design of Flexible Structured Fuzzy Controller

6.2.1 A Flexible Structured Fuzzy Controller

In designing of a fuzzy controller, several factors besides the fuzzy rules need to be determined by the designer and thus there is much room for optimization. Such factors include the T-norm, the defuzzification method and the scaling factors. In [247], a fuzzy controller with a flexible structure has been proposed. The structure of the fuzzy controller is flexible because several operations, including the fuzzy intersection and the defuzzification, are parameterized, which are subject to adaptation. In the fuzzy controller, a so-called soft T-norm has been employed:

$$\tilde{T}(x,y) = (1-\alpha)\frac{1}{2}(x+y) + \alpha T(x,y), \qquad (6.1)$$

where $T(x,y)$ is a conventional T-norm, $\alpha \in [0,1]$ is a parameter to be adjusted. Thus, given a set of fuzzy control rules:

R^i : If $e(k)$ is $A^i(e)$ and $\Delta e(k)$ is $B^i(\Delta e)$, then Δu is $C^i(\Delta u), i = 1, \cdots, N,$

$$(6.2)$$

where,

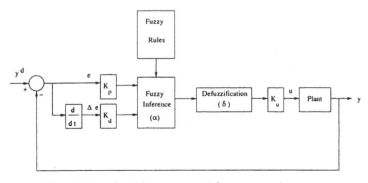

Fig. 6.2. A flexible structured fuzzy control system.

$$e(k) = K_p(y^d(k) - y(k)), \tag{6.3}$$
$$\Delta e(k) = K_d(e(k) - e(k-1)), \tag{6.4}$$

where K_p and K_d are two scaling factors, $\Delta u(k)$ is the change of the control, A^i, B^i, and C^i are fuzzy subsets defined on the universe of the corresponding variable. Similar to the Mamdani fuzzy controller, the output of the fuzzy controller is given by:

$$\Delta u = \frac{\sum_{i=1}^{N} (w^i)^\delta \cdot u^i}{\sum_{i=1}^{N} (w^i)^\delta}, \tag{6.5}$$
$$w^i = \tilde{T}(\mu_{A^i}(e), \mu_{B^i}(\Delta e)), \tag{6.6}$$

where u^i is the center of fuzzy set C^i, $\delta \in [1, \infty)$. It can be seen that when $\delta = 1$, the defuzzification is the center of area (COA) method, and when $\delta \to \infty$, the defuzzification is reduced to the mean of maximum (MOM) method. Finally, the output of the controller is as follows:

$$u(k) = u(k-1) + K_u \Delta u. \tag{6.7}$$

The structure of the fuzzy controller is illustrated in Fig. 6.2.

6.2.2 Parameter Optimization Using Evolution Strategies

The standard evolution strategy (ES) is used for the optimization of the parameters in the flexible structured fuzzy controller. The chromosome of the ES has 10 genes, 5 of which are the parameters to be optimized, and the other 5 genes are known as strategy parameters, see Fig. 6.3.

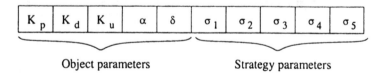

Fig. 6.3. A chromosome of the ES.

The target of optimization is to obtain a fuzzy controller that has high transient and static control performance. Therefore, the following cost function is to be minimized:

$$J = \lambda_1 t_r + \lambda_2 \sum_{k=0}^{\infty} |e(k)|, \tag{6.8}$$

where t_r is the rise time of the system's step response, λ_1 and λ_2 are two positive constants. In addition, the constraints on the parameters are:

$$K_p, K_d, K_u \quad > \quad 0, \tag{6.9}$$
$$0 \leq \alpha \leq 1, \tag{6.10}$$
$$\delta \quad \geq \quad 1. \tag{6.11}$$

6.2.3 Simulation Study

In the simulation, the parent population size is set to 15, and the the offspring population size is set to 100, and the optimization is run for 500 generations.

One linear and one nonlinear systems are considered in the simulation. The linear system is a second-order non-minimum phase system that can be described by the following transfer function:

$$G(z) = \frac{-0.073z^{-1} + 0.094z^{-2}}{1.0 - 1.683z^{-1} + 0.705z^{-2}}. \tag{6.12}$$

The nonlinear system used in the simulation is:

$$y(k+1) = \frac{y(k)y(k-1)(y(k)+2.5)}{1.0 + y^2(k) + y^2(k+1)} + 1.3\sin(y(k)+1.0)u(k). \tag{6.13}$$

The unit step response is the desired output of the systems. To investigate the performance of different T-norms, the following four T-norms are considered:

$$T_1(x,y) = \min(x,y), \tag{6.14}$$
$$T_2(x,y) = xy, \tag{6.15}$$
$$T_3(x,y) = \max(x+y-1,1), \tag{6.16}$$
$$T_4(x,y) = \frac{xy}{x+y-xy}. \tag{6.17}$$

Without the loss of generality, error, change of error and change of control are partitioned into 7 fuzzy subsets, namely, {NB, NM, NS, ZO, PS,PM,PB}. The fuzzy rules are obtained heuristically as shown in Fig. 6.4.

The three scaling factors are initialized to 1.0, $\alpha = 1$ (T-norm only), and $\delta = 1$ (COA defuzzification). The response of the linear and nonlinear systems are shown in Fig. 6.5, Fig. 6.6 and Fig. 6.7, Fig. 6.8, respectively. It can be seen that the four T-norms exhibit very similar performance on the linear system. In contrast, the most widely used minimum operation has best performance on the nonlinear system.

The parameters are then optimized with the ES. The system responses under the control of the optimized fuzzy control are presented in Fig. 6.9, Fig. 6.10, Fig. 6.11 and Fig. 6.12.

Some general remarks can be made from the optimization results. Obviously, the optimization of the parameters of the flexible structure fuzzy controller is able to improve the performance significantly, no matter which T-norm is used for fuzzy intersection. Meanwhile, different T-norms are comparable in performance, once they are optimized.

E \ CE ΔU	NB	NM	NS	ZO	PS	PM	PB
NB	PB	PB	PB	PB	PM	PS	ZO
NM	PB	PB	PM	PM	PS	ZO	ZO
NS	PB	PM	PM	PS	ZO	ZO	NS
ZO	PM	PS	PS	ZO	NS	NS	NM
PS	PS	ZO	ZO	NS	NM	NM	NB
PM	ZO	ZO	NS	NM	NM	NB	NB
PB	ZO	NS	NM	NB	NB	NB	NB

Fig. 6.4. The fuzzy control rule table.

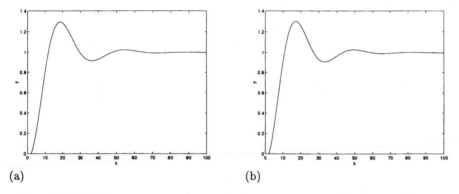

(a)   (b)

Fig. 6.5. The linear system response before optimization. (a) T_1 and (b) T_2.

6.3 Evolutionary Optimization of Fuzzy Rules

6.3.1 Genetic Coding of Fuzzy Systems

Genetic representation of the fuzzy system to be optimized is essential to the optimization result. Before the fuzzy system is encoded, it needs to be determined which evolutionary algorithm should be used. Note that the parameter and structure optimization of fuzzy systems is an optimization problem with mixed data types. Generally, evolution strategies are well suitable for real parameter optimizations. If an ES is used for both parameter and structure optimization, the hybrid ES that is able to deal with real and integer numbers developed in [4] should be employed. Assume a fuzzy membership A_{ij} consists of two parameters: center (c_{ij}) and width (w_{ij}), where $i = 1, 2, ..., n$ is the number of input variables, and $j = 1, 2, ..., m_i$ is the number of fuzzy

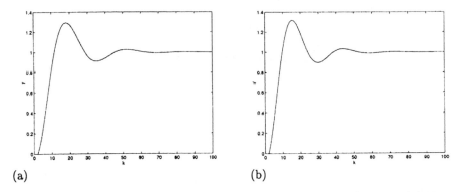

Fig. 6.6. The linear system response before optimization. (a) T_3 and (b) T_4.

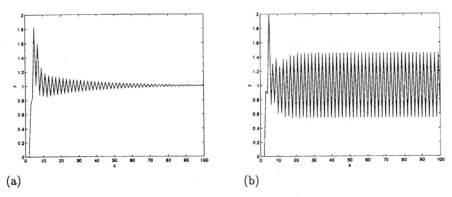

Fig. 6.7. The nonlinear system response before optimization. (a) T_1 and (b) T_2.

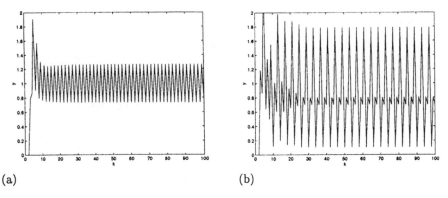

Fig. 6.8. The nonlinear system response before optimization. (a) T_3 and (b) T_4.

subsets in the fuzzy partition of variable x_i, and $(c_k, w_k), k = 1, ..., m$ are the center and width of output membership function B_k, then a chromosome for the parameter and structure optimization using ES is illustrated in Fig. 6.13. In rule structure coding, s_{ij} is an integer that satisfies $0 \leq s_{ji} \leq m_i$. If

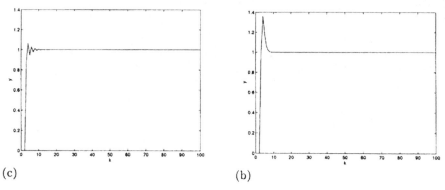

(c)   (b)

Fig. 6.9. The linear system response after optimization. (a) T_1 and (b) T_2.

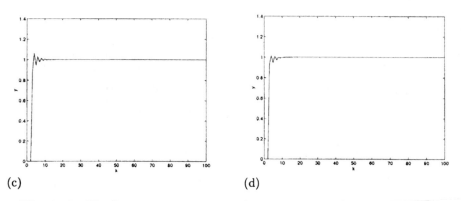

(c)   (d)

Fig. 6.10. The linear system response after optimization. (a) T_3 and (b) T_4.

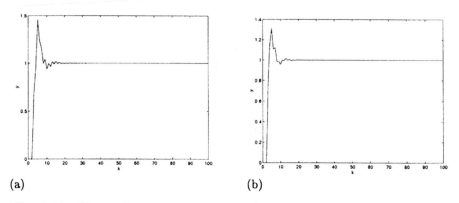

(a)   (b)

Fig. 6.11. The nonlinear system response after optimization. (a) T_1 and (b) T_2.

$s_{ji} = 0$, then the i-th input variable does not appear in the i-th rule. Note that the strategy parameters are not shown in the chromosome. In contrast, if a genetic algorithm is employed, all parameters need to be converted into a binary string. The string length of the membership parameters needs to

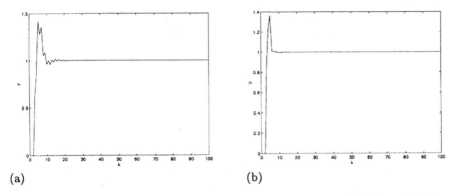

(a)   (b)

Fig. 6.12. The nonlinear system response after optimization. (a) T_3 and (b) T_4.

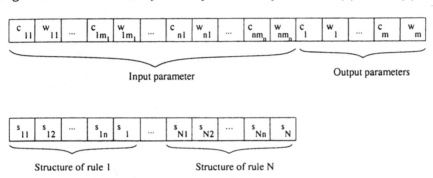

Fig. 6.13. Coding of a fuzzy system using an evolution strategy.

be fixed according to the range of the variables. The string length of the structure parameters is easy to determine. For example, if we assume that the maximal number of fuzzy subsets in a fuzzy partition is 7, then 3 bits are needed for each structure parameter. A larger number of fuzzy subsets is not desirable because a too large number of fuzzy subsets will harm interpretability of fuzzy systems.

Genetic programming can also be employed for optimization of fuzzy systems. Suppose we have the following two fuzzy rules:

If x_1 is A_{11} and x_2 is A_{21}, then y is B_1;
If x_1 is A_{12} and x_2 is A_{22}, then y is B_2.

The genetic representation of the two fuzzy rules is presented in Fig. 6.14. It can be seen that the function set for a fuzzy system is as follows:

$$F = \{\text{ANT}, \text{CONS}, f_{AND}, \text{IF-THEN}, f_{OR}\}, \tag{6.18}$$

where "ANT" returns the degree of membership for the premise variable, and "CONS" the degree of membership for the consequent variable. If the TSK fuzzy rules are to be generated, "CONS" can be a function of the input variables. "f_{AND}" is the fuzzy intersection operator, and "f_{OR}" is the fuzzy union. A_{ij} and B_k are fuzzy sets defined on the universe of discourse of the corresponding variables.

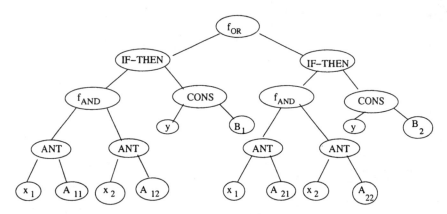

Fig. 6.14. Coding of the two rules using a genetic programming.

6.3.2 Fitness Function

The fitness function in evolutionary fuzzy systems design can be very different, depending on the main purpose of the fuzzy system. If the fuzzy system is a fuzzy controller, the evolutionary algorithm is then to minimize the regulation or tracking error. Actually, the TSK fuzzy rules have found more successful applications in constructing fuzzy controllers. Usually, the integral of the squared error (ISE), the integral of the absolute error (IAE), the integral of time by squared error (ITSE) and the integral of time weighted absolute error (ITAE) can be used:

$$J_{ISE} = \int_{t_1}^{t_2} (y(t) - y^d(t))^2 dt, \tag{6.19}$$

$$J_{IAE} = \int_{t_1}^{t_2} |y(t) - y^d(t)| dt, \tag{6.20}$$

$$J_{ITSE} = \int_{t_1}^{t_2} t(y(t) - y^d(t))^2 dt, \tag{6.21}$$

$$J_{ITAE} = \int_{t_1}^{t_2} t|y(t) - y^d(t)| dt, \tag{6.22}$$

where y and y^d are the output and the desired output of the system. The different performance criteria have different focuses. For example, the ISE strongly suppress large errors, while the IAE strongly suppress large errors such as large overshoot, and the ITAE strongly penalizes persisting errors. The different measures can also be combined to achieve a trade-off among different performances.

Other objectives, such as noise sensitivity and robustness have often to be considered in design of a fuzzy controller. For example, the linear quadratic Gaussian (LQG) control is able to reach a compromise between the control error and the control variance that is caused by the presence of noise:

$$J = \lim_{k \to \infty} E\{e^2(t) + \lambda u^2(t)\}, \tag{6.23}$$

$$e(t) = y(t) - y^d(t). \tag{6.24}$$

Besides, control objectives such as H_2 and H_∞ performance can also be considered. Generally speaking, the H_2 performance is useful to specify desired performance requirements on the controller in the presence of measurement noise and random disturbance, whereas the H_∞ performance guarantees the robustness to model uncertainty.

If the fuzzy system is generated for system modeling, the model accuracy is the most popular performance index. Nevertheless, other objectives, especially the interpretability considerations, such as the rule complexity, distinguishability of the fuzzy partitions, and the consistency of the fuzzy rules are also very important. The objectives for the interpretability of the fuzzy systems may be very helpful in gaining an insight into unknown systems.

6.3.3 Evolutionary Fuzzy Modeling of Robot Dynamics

Decentralized Fuzzy Modeling of A Robot Manipulator. It is very often necessary to calculate the robot dynamics in robot control. Two main approaches are used by most researchers to derive the dynamic model of a manipulator – the Lagrange-Euler(L-E) and the Newton-Euler(N-E) formulations. From the control point of view, the L-E formulation is very desirable. For an n-link robot arm, the Lagrange equation of motion is as follows:

$$\tau_i = \sum_{j=1}^{n} D_{ij}(q)\ddot{q}_j + \sum_{j=1}^{n}\sum_{k=1}^{n} C_{ijk}(q)\dot{q}_j\dot{q}_k + G_i(q), \ i = 1, ..., n, \tag{6.25}$$

where τ_i is the torque exerted on joint i, q_i is i-th joint displacement, D_{ii} and D_{ij} are the effective inertia and coupling inertia, C_{ijk} stands for the centrifugal and Coriolis forces, and G_i is the gravity loading. Note that for a planned trajectory, the desired torque depends not only on the trajectory, the geometric and inertia parameters of the link itself, but also on the parameters of other links and the payload at the end effector.

In order to model the dynamics of each link with a fuzzy system, it is necessary to choose proper input and output variables. For the sake of computational simplicity, it is necessary and feasible to choose a decoupled fuzzy system. In our case, only position and velocity are chosen as two input variables and, naturally the feedforward torque is chosen as the output. Consequently, the fuzzy rules in the feedforward system can be expressed in the following form:

$$\text{If } q^d(k) \text{ is } A_1^i \text{ and } \dot{q}^d(k) \text{ is } A_2^i, \text{ then } \tau \text{ is } \tau_0^i,$$

where A_1^i and A_2^i are the fuzzy sets for q^d and $\dot{q}^d$, τ_0^i is the crisp output of each fuzzy rule and k is the time instant. Note that the zero-order TSK fuzzy rule is used, because it is found that higher order TSK fuzzy rules can not improve the accuracy of the fuzzy model much. Accordingly, if the rule base has M rules, the final output of the fuzzy model is calculated as follows:

$$\tau(k) = \frac{\sum_{i=1}^{M}\{w^i(k)\tau_0^i\}}{\sum_{i=1}^{M} w^i(k)}, \tag{6.26}$$

$$w^i(k) = \min\{A_1^i(q^d(k)), A_2^i(\dot{q}^d(k))\}. \tag{6.27}$$

A genetic algorithm is employed to optimize the parameter and structure of the fuzzy rules. Two additional measures are introduced to the canonical GA to improve the convergence of the algorithm. First, we stochastically introduce a randomly generated chromosome at a probability of R_h to replace one of the two parents selected for reproduction. Second, select the best performed genes in the current population at a rate of R_e and place them directly in the next generation. This is similar to the elitism strategy, however, it is not deterministic.

These two measures can be further explained with the results in [191]. If the reproduction is carried out in the traditional way, the best gene will probably get lost and thus the convergence of the search algorithm can not be guaranteed. However, if we adopt the second measure only, the procedure of evolution is no longer a Markov process and thus it does not satisfy the assumption of the convergence theorem. Despite our successful application of these measures, mathematical analyses of them still lack. The parameters R_h and R_e are adjusted in the following way. At the beginning of learning, R_h is relatively large and R_e is small. As the evolution proceeds, R_h is gradually decreased and R_e is increased.

In order that the feedforward fuzzy system can learn the mapping of the robot inverse dynamics, the following quadratic form of performance index is established:

$$J = \sum_{k=0}^{P}(\tau^d(k) - \tau(k))^2, \tag{6.28}$$

where $\tau^d(k)$ and $\tau(k)$ are the desired torque and torque computed from the feedforward fuzzy system, respectively, k is the time index, P is the number of the training samples. Because the genetic algorithm maximizes the fitness function, and because our aim is to minimize the above performance index, the fitness function of each gene is calculated as follows:

$$F = \frac{1}{1 + J},\tag{6.29}$$

where J is the performance index and 1 is introduced to avoid computational problems. Suppose the membership functions in the fuzzy system take a Gaussian form as:

$$A_j^i(x) = \exp(-b_j^i(x - a_j^i)^2), \; j = 1, 2,\tag{6.30}$$

where a_j^i and b_j^i are the center and the width of the Gaussian function. For simplicity, the membership function in equation (6.30) is notated as (a_j^i, b_j^i).

The coding of the parameters to be adjusted can be arranged as in Fig. 6.15. As usual, the Pittsburg approach is adopted to avoid the fitness assignment problem. In this study, each input variable is partitioned into n fuzzy subspaces, thus, the number of rules $M = n^2$ if the grid structure is used. In the coding, the length of binary bits for each parameter is manually fixed. The basic principle is that the length of the bits should be able to represent the possible value of the parameter to a sufficient accuracy. However, an overly large number of bits will unnecessarily increase the length of the entire chromosome, which may make the search inefficient. Therefore, each parameter to be optimized can be normalized to a certain range.

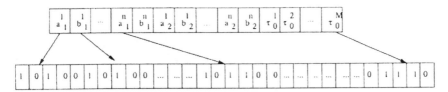

Fig. 6.15. Genetic coding of the rule parameters of the fuzzy system.

The following observations can be made concerning the optimization of fuzzy rules using a genetic algorithm:

- The search of genetic algorithms starts from multiple initial states simultaneously and proceeds in all of the parameter sub-spaces in parallel, which provides the GA an excellent parallel processing ability and an inherent global optimization capacity.
- The GA requires almost no priori knowledge of the system involved in optimization, which enables the GA to deal with completely unknown systems that other optimization methods fail.
- The GA can not evaluate the performance of a system properly at one step. For this reason, it can generally not be used as an on-line optimization strategy and is generally more suitable for fuzzy modeling than for fuzzy control.

Without the loss of generality, we take a two-link rigid robot as an example for simulation. It should be pointed out that in training the fuzzy system, the algorithm does not require full knowledge of the robot inverse model because the optimization is completely data-driven. In practice, the training data can be collected by carrying out experiments or by establishing an ideal mathematical model. If the data are collected by experimentation, we can exert a bounded random torque on the robot to be controlled. It may also be possible to derive an ideal mathematical model. This is theoretically feasible and helpful for training and checking of the fuzzy system, despite that there are uncertainties in the mathematical model. In computer simulation, we need a model to simulate the behavior of a robot to collect the training data. The robot model used in simulation is to be given later, where $[m_1, m_2]^T = [2, 1]^T$, $[l_1, l_2]^T = [0.2230.2]^T$. During the training, no nonlinear friction and payloads are considered. The trajectory of the two links is defined as follows:

$$q_1^d(k) = 0.5\pi(1 - e^{-k}) \text{ (rad)}, \tag{6.31}$$
$$q_2^d(k) = \pi(1 - e^{-k}) \text{ (rad)}. \tag{6.32}$$

First of all, both input variables in each link are partitioned into four fuzzy subsets and thus 16 fuzzy rules will be generated for each link. Then, genetic algorithm is used to tune the parameters so that the fuzzy system can realize the mapping of the inverse robot dynamics. The population size of the GA is 50 and the length of each gene is 160. The crossover and mutation probability are set to 0.95 and 0.1, respectively. After about 60 generations of evolution, the GA search algorithm is convergent. The termination criterion is given by an average mapping error determined beforehand. The fuzzy models for link 1 and link 2 resulted from the best chromosome are shown in Table 6.1 and Table 6.2, respectively. For example, the first fuzzy rule in Table 6.1 is:

If q^d is $(0, 3.85)$ and $\dot{q}^d$ is $(0.0625, 5.85)$, then τ is 4.375.

Fig. 6.16 shows the approximation results of the fuzzy systems. The average approximation errors are 0.134 and 0.082, respectively.

If the performance of a fuzzy system is only evaluated by the approximation accuracy, the above fuzzy systems with a grid structure are acceptable. However, we find in simulation that the average firing rates of the rules are very low. For example, on average only 50% and 22% of fuzzy rules in the rule base of link 1 and link 2 are fired at each time instant. It indicates that the fuzzy systems are not compact enough and the structure of the fuzzy rules needs to be optimized.

It is straightforward to optimize the structure and parameters of the fuzzy rules simultaneously using genetic algorithms. Each fuzzy system is represented as a string composed of two sub-strings. The first sub-string, which has the same form as illustrated in Fig. 6.15, is to optimize the parameters of the fuzzy systems. The second sub-string encodes the structure of the fuzzy rule such that one integer number represents one membership function in the space of the input variable in question. The membership functions in the first sub-string are numbered in ascending order according to the value of

Table 6.1. Rule base for link 1.

IF		THEN
q^d	$\dot{q}^d$	τ_0^i
(0, 3.85)	(0.0625, 5.85)	4.375
(1.06, 4.35)	(0.0625, 5.85)	2.75
(0, 3.85)	(0.25, 1.35)	6.5
(1.06, 4.35)	(0.25, 1.35)	5.75
(0, 3.85)	(1.0625, 5.6)	2.375
(1.06, 4.35)	(1.0625, 5.6)	5.625
(0, 3.85)	(1.4375, 7.6)	-0.875
(1.06, 4.35)	(1.4375, 7.6)	2.75
(0.25, 3.6)	(0.0625, 5.85)	2.5
(1.0, 3.6)	(0.0625, 5.85)	4.125
(0.25, 3.6)	(0.25, 1.35)	6.5
(1.0, 3.6)	(0.25, 1.35)	6.875
(0.25, 3.6)	(1.0625, 5.6)	3.375
(1.0, 3.6)	(1.0625, 5.6)	6.625
(0.25, 3.6)	(1.4375, 7.6)	-0.75
(1.0, 3.6)	(1.4375, 7.6)	5.0

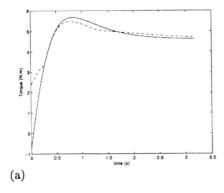

(a)

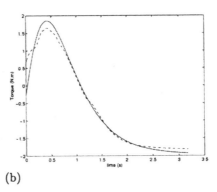

(b)

Fig. 6.16. Off-line training of the inverse dynamics (without structure optimization). (a) Link1; (b) Link 2. Solid line: the desired torque; dashed line: the fuzzy system approximation.

the center. For example, a number '1' represents the membership function with the lowest center. Since each variable is supposed to have at most 4 subspaces, the valid numbers in the second sub-string are '0', '1', '2', '3', and '4'. The number '0' implies that this variable does not appear in the premise part of the rule. If both variables take a value of '0' in the second sub-string,

Table 6.2. Rule base for link 2.

IF		THEN
q^d	$\dot{q}^d$	τ_0^i
(0.375, 5.85)	(0.219, 2.35)	0.75
(2.25, 7.6)	(0.219, 2.35)	-0.6875
(0.375, 5.85)	(1.25, 5.1)	0.625
(2.25, 7.6)	(1.25, 5.1)	1.4375
(0.375, 5.85)	(1.875, 6.85)	1.5
(2.25, 7.6)	(1.875, 6.85)	1.6825
(0.375, 5.85)	(2.843, 2.35)	0.875
(2.25, 7.6)	(2.843, 2.35)	0.5
(1.0. 4.6)	(0.219, 2.35)	-1.25
(2.843, 2.35)	(0.219, 2.35)	-1.9375
(1.0. 4.6)	(1.25, 5.1)	1.5
(2.843, 2.35)	(1.25, 5.1)	-1.5
(1.0. 4.6)	(1.875, 6.85)	1.75
(2.843, 2.35)	(1.875, 6.85)	1.6875
(1.0. 4.6)	(2.843, 2.35)	1.5
(2.843, 2.35)	(2.843, 2.35)	-1.4375

then this rule is deleted from the rule base. It is also possible that more than one rule in the rule base has the same premise. In this case, only the rule that appears first is kept, so that the rules are consistent. An example of the second sub-string is given Fig. 6.17. The corresponding fuzzy rules are:

$$R^1: \text{If } q^d(k) \text{ is } (a_1^3, b_1^3) \text{ and } \dot{q}^d(k) \text{ is } (a_1^4, b_1^4), \text{ then } \tau \text{ is } \tau_0^1;$$
$$R^2: \text{If } \dot{q}^d(k) \text{ is } (a_2^1, b_2^1), \text{ then is } \tau \text{ is } \tau_0^2;$$
$$\vdots$$
$$R^{16}: \text{(Deleted)}$$

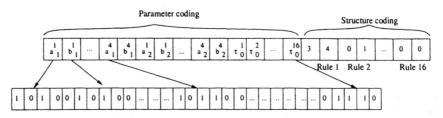

Fig. 6.17. Genetic coding of the parameters and the structure of the fuzzy system.

In order to optimize the fuzzy rule structure, the performance index in equation (6.28) is rewritten as:

$$J = \sum_{k=0}^{P} [\tau^d(k) - \tau(k)]^2 + \lambda J_{COM}, \tag{6.33}$$

where λ is a weighting constant, J_{COM} is the penalty for model complexity and is expressed as:

$$J_{COM} = \frac{\text{total number of rules}}{\text{average number of fired rules}}. \tag{6.34}$$

The value of λ is set to 0.1 for link 1 and 0.25 for link 2. We suppose a rule is fired when the truth value in equation (6.27) is greater than 0.05. In case no rules are fired or no rules are left in the rule base, J_{COM} will be set to a very large value.

The simulation results are encouraging. The optimized rule bases for link 1 and link 2 have 8 and 11 rules, respectively and the firing rates are raised to about 75% and 32%, respectively. The rule bases for the two links are listed in Tables 6.3 and 6.4 respectively, and the approximation results are presented in Fig. 6.18. The average approximation errors are 0.132 and 0.122, respectively. It can be seen that the approximation performance is quite satisfactory, although the rules in the base are reduced.

Table 6.3. Rule base for link 1 (structurally optimized).

IF		THEN
q^d	$\dot{q}^d$	τ_0^i
(1.75, 5.85)	(0.375, 0.35)	6.875
(1.375, 5.85)	(0.125, 5.6)	6.25
(1.0625, 5.85)	(0.125, 5.6)	6.375
(0.4375, 5.1)	(0.125, 5.6)	1.5
(0.4375, 5.1)	(1.0, 7.85)	-0.125
(0.4375, 5.1)	(1.25, 4.1)	5.75
(1.0625, 5.85)	(0.375, 0.35)	6.75
(0.4375, 5.1)		6.625

Table 6.4. Rule base for link 2 (structurally optimized).

IF		THEN
q^d	$\dot{q}^d$	τ_0^i
(2.8125, 4.35)	(0.1563, 5.6)	-1.875
(0.2188, 7.35)	(1.375, 1.85)	-0.0625
(1.4375, 2.1)	(1.375, 1.85)	1.375
(2.8125, 4.35)	(1.9375, 3.6)	0.4375
(1.625, 5.1)	(2.5313, 1.35)	-1.75
(2.8125, 4.35)	(2.5313, 1.35)	0.8125
(1.625, 5.1)	(1.9375, 3.6)	-1.25
(1.4375, 2.1)	(1.9375, 3.6)	-1.5
(1.4375, 2.1)	(0.1563, 1.65)	1.625
(1.4375, 2.1)		-1.25
	(0.1563,1.65)	0.5625

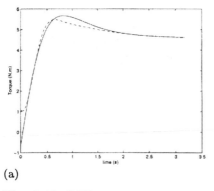

(a)

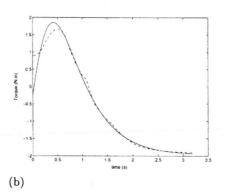

(b)

Fig. 6.18. Off-line training of the inverse dynamics (with structure optimization). (a) Link1; (b) Link 2. Solid line: the desired torque; dashed line: the fuzzy system approximation.

6.4 Fuzzy Systems Design for High-Dimensional Systems

6.4.1 Curse of Dimensionality

Curse of dimensionality [11] refers to the exponential growth of hypervolume as a function of dimensionality. The main problem with regard to the curse of dimensionality in fuzzy systems is that the number of fuzzy rules increases exponentially with the number of inputs of the fuzzy system. For example, if there are n input variables, and each input variable is partitioned into m fuzzy subsets, then m^n fuzzy rules are needed to cover the whole input space if a grid structure is used. In evolutionary fuzzy system generation, this makes the evolutionary algorithm very inefficient. If the the coding in Fig. 6.17 is used for a fuzzy system with m^n fuzzy rules, then a total number of $(3n+1) \times m^n$ parameters are needed. For example, if $n = 10$ and $m = 3$, the number of parameters will amount to 10×3^{10}, which makes the optimization problem intractable.

A huge number of fuzzy rules not only increases the computational complexity of the fuzzy system, but also harms the interpretability and thus makes the fuzzy system incomprehensible.

In addition, the curse of dimensionality also gives rise to the problem that a variety of training data are required for the generation of the fuzzy rules. In the following sections, a number of methods will be introduced to tackle the first aspect of the curse of dimensionality in fuzzy systems, that is, to reduce the number of fuzzy rules.

6.4.2 Flexible Fuzzy Partitions

In fuzzy systems generation, the grid structure is the most widely used, see Fig. 6.19. Although this structure has the advantage of easy understandability on the one hand, it is suffered from the curse of dimensionality on the other hand.

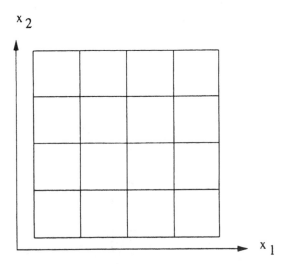

Fig. 6.19. The grid structure of a fuzzy system. For a two-dimensional system, 16 fuzzy rules are needed if each parameter is partitioned into 4 subsets.

To overcome this problem, more flexible, non-grid structures, such as k-d tree (Fig. 6.20) and the quad tree (Fig. 6.21) can be used to partition the input space of a fuzzy system.

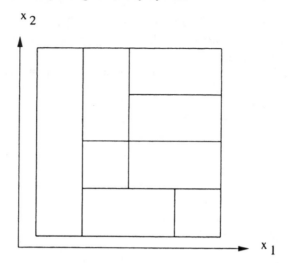

Fig. 6.20. The k-d tree structure of a fuzzy system. For a two-dimensional system, only 8 fuzzy rules are needed for the given partition, although each parameter is still partitioned into 4 subsets.

A weakness of the k-d tree partition is that the interpretability of the fuzzy subsets may be degraded. An alternative to the k-d tree structure is the q-d tree structure, as shown in Fig. 6.21, which was proposed in [212]. The basic idea is that the input space is first partitioned roughly. When

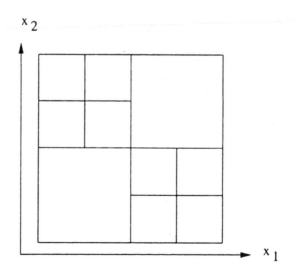

Fig. 6.21. The q-d tree structure of a fuzzy system. For a two-dimensional system, 10 fuzzy rules are needed for the given partition. Each parameter is partitioned into 4 subsets.

necessary, a subspace is recursively divided until the approximation quality is satisfactory.

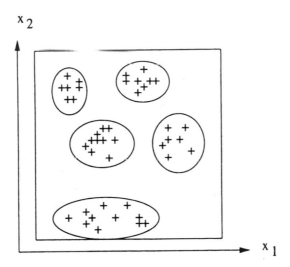

Fig. 6.22. Partition of the input space by clustering.

A more flexible, data-driven partition of the input space is to use clustering methods to determine the rule structure. See Fig. 6.22 for an illustrative example. However, similar to the k-d tree partition, the interpretability of the fuzzy subsets may be difficult.

6.4.3 Hierarchical Structures

An efficient method for reducing the number of fuzzy rules is to use more than one hierarchy in the rule structure. Fig. 6.23 shows four examples of such hierarchical structures. It can be seen that there will be a trade-off between the number of layers and the number of rules. Meanwhile, the structure of the fuzzy system can vary a lot.

It has been found that hierarchical fuzzy systems do not necessarily have the same functional approximation capability as the non-hierarchical one. This is quite obvious because the number of mappings which can be represented by a hierarchical fuzzy system is less than that of a non-hierarchical one. It has been shown that a hierarchical fuzzy model is able to represent any function only if it is functional complete [137], which means that the hierarchical expansion is an orthogonal expansion of the original model. Refer to [137] for detailed discussions.

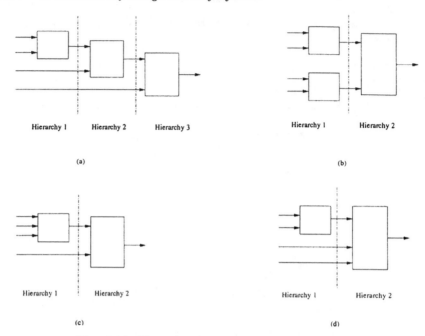

Fig. 6.23. Hierarchical structures of fuzzy systems.

6.4.4 Input Dimension Reduction

A more direct method for dealing with the curse of dimensionality is to reduce the input dimension. This is feasible under the assumption that either some of the inputs are irrelevant to the output, or some of the inputs are correlated with each other. In the former case, the irrelevant inputs can be dropped so that the input dimension can be reduced. In the later case, the correlated input vector is transformed into a zero-mean, uncorrelated vector of a lower dimensionality. For these purposes, mutual information and the principal component analysis can be used.

Mutual Information. The concept of mutual information was originated from the information theory [41]. Mutual information can be seen as a reduction of entropy in Y given the X:

$$MI(X;Y) = H(Y) - H(Y|X), \tag{6.35}$$

where $H(Y)$ is the entropy of Y and $H(Y|X)$ is the conditional entropy. It is obvious that the following equation also holds:

$$MI(X;Y) = H(X) - H(X|Y). \tag{6.36}$$

From statistics point of view, it is a measure for the statistical independence between the two random variables (X, Y). In other words, mutual information is a measure of statistical correlation between X and Y.

Assume $p(x)$ and $p(y)$ are the probability density of X and Y, $p(x, y)$ is the joint probability of X and Y, the mutual information (MI) of two random variables X and Y is defined as follows:

$$MI(X;Y) = \sum_{x_i} \sum_{y_i} p(x_i, y_i) \log \frac{p(x_i, y_i)}{p(x_i)p(y_i)}. \tag{6.37}$$

The physical meanings of the definition are quite clear. $MI(X;Y)$ equals zero if and only if X and Y are independent. Furthermore, the following equation holds for mutual information:

$$MI(X;Y) = MI(Y;X). \tag{6.38}$$

In designing fuzzy systems, mutual information can be used to select a few inputs that are most relevant to the output, so that the input dimension can be reduced. To calculate the mutual information, it is necessary to estimate the probability density and the joint probability of the inputs and the output, which is not trivial in many cases. There are several methods for calculating probability [18], among which the histogram is the simplest non-parametric method that is most widely used.

Basically, the histogram method divides the sample space into bins and counts samples in each bin. For a set of samples observed from a random variable x, and suppose x_0 is minimal value of the samples, the bins of the histogram are defined to be the intervals $x_i \in [x_0 + (i-1)h, x_0 + ih)$, where $= 1, ..., m$, m is the number of bins, h is the bin width. The intervals have been chosen closed on the left and open on the right for definiteness, see Fig. 6.24. Thus, the probability density can be estimated in the following way:

$$p(x_i) = \frac{N_i}{N}, \tag{6.39}$$

where N_i is the number of samples in the interval x_i, N is the number of samples in total. Thus, an estimation of the true density can be obtained.

Usually, the accuracy of the estimation depends on the width of the bins. The larger the bin width, the lower the resolution is. However, if small bins are used, many bins may be empty, which means that no data samples are in the bins. This problem becomes serious, especially when the sample space is sparse and the dimension is high.

As an example, we consider a simulated data sample with 11 inputs and one output [114]. The data set consists of 20000 training data and 8000 test data. The training data are used for estimating the mutual information and the results are listed in Table 6.5.

From the table, it can be seen that x_5, x_4 and x_9 are the most relevant inputs to the output according to the mutual information measure. Using the direct method for fuzzy rule generation introduced in Chapter 4, a fuzzy system is obtained with 16 fuzzy rules. The root mean square error (RMSE) on the training data is 0.309 and on the test data is 0.318.

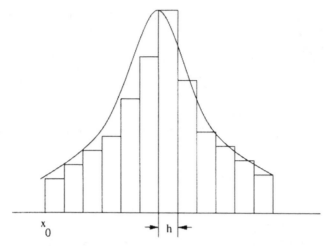

Fig. 6.24. Probability density estimation using histogram.

Table 6.5. Mutual information between the inputs and the output.

X_i	$MI(X_i, Y)$
X_1	0.046
X_2	0.068
X_3	0.086
X_4	0.224
X_5	0.566
X_6	0.020
X_7	0.050
X_8	0.012
X_9	0.106
X_{10}	0.026
X_{11}	0.009

To show that mutual information is a proper measure for input selection, a fuzzy system with all the inputs are also generated using the same method. The obtained rule base contains 1233 fuzzy rules, and the RMSE on the training data is 0.259 and on the test data is 0.305. It can be seen that although the training error is smaller than the fuzzy system with a reduced dimension of 3 inputs, the test error is very similar. Note that the number of fuzzy rules of the simplified system is much smaller than the full system.

Finally, the three inputs that have the minimal mutual information value, namely, x_6, x_8 and x_{11} are then used to generate a fuzzy system using the same method. As expected, the quality of the fuzzy system becomes much worse. The achieved RMSE on the training data is 0.657 and on the test data is 0.653, both of which are significantly larger than the full system.

Principal Component Analysis. Principal component analysis (PCA) is a mathematical procedure that transforms a number of possibly correlated variables into a smaller number of uncorrelated variables called principal components. The first principal component accounts for as much of the variability in the data as possible, and each succeeding component accounts for as much of the remaining variability as possible.

Suppose we have p samples for an n-dimensional vector. Denote the samples as X of dimension $n \times p$. First of all, the matrix needs to be standardized so that it is zero mean with variance of 1. Denote the new matrix with $Z = \{z_i, i = 1, .., p\}$. The goal of transformation is to find a linear function y for each z_i so that the elements of y are uncorrelated and the mean is zero:

$$y = V^T z_i, V = [v_1, ..., v_p]. \tag{6.40}$$

There are an infinite number of values for V that will satisfy equation (6.40). A criterion to get a unique V is to maximize the variance of y. In other words, for the first principal component v_1, a solution is obtained by maximizing the variance of y_1:

$$\text{Maximize } \frac{1}{n} \sum_{i=1}^{n} y_{1i}^2 = v_1^T R v_1, \tag{6.41}$$

where, R is the correlation matrix of z_1, and

$$y_{1i} = v_1^T \cdot z_i, \ z_i : n \times 1, \tag{6.42}$$

subject to

$$v_1^T v_1 = 1, \ v_1 : 1 \times n. \tag{6.43}$$

Using the Lagrange method:

$$\phi_1 = v_1^T R v_1 - \lambda_1 (v_1^T v_1 - I), \tag{6.44}$$

where λ_1 is the eigenvalue of the first principal component, and let

$$\frac{\partial \phi_1}{\partial v_1} = 0, \tag{6.45}$$

we have:

$$(R - \lambda_1 I) v_1 = 0, \tag{6.46}$$

where I is the unit matrix. The second principal component can be obtained by maximizing the variance of $v_2^T z_i$ subject to

$$v_2^T v_2 = 1, \tag{6.47}$$

$$\text{Cov}(v_1^T z_i, v_2^T z_i) = 0. \tag{6.48}$$

Further principal components can be obtained in a similar way.

The needed number of principal components should be determined by checking the eigenvalue of the principal components. The eigenvalues for each principal component corresponds to the amount of total variance in the data described by this component. A principal component is significant if its eigenvalue is greater than the corresponding value for the case of pure noise.

Principal components analysis can be performed by a linear neural network, whose input and output are the inputs of the original data and the hidden nodes are the principal components, which can then be used for constructing fuzzy rules, see Fig. 6.25. Obviously, the number of hidden nodes (m) should be smaller than n to realize dimension reduction.

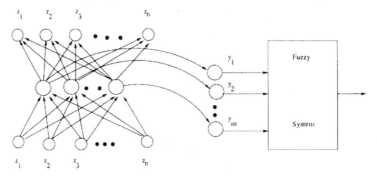

Fig. 6.25. A linear neural network is trained to map the inputs onto themselves. Such a linear network is equivalent to the principal component analysis, where the hidden nodes represent the principal components.

To illustrate this method, a linear neural network is used to perform principal component analysis on a data set obtained in aerodynamic design optimization [128]. The data consists of 1500 samples with 52 inputs and one output. The 52 inputs are the geometrical coordinates describing a turbine blade. The number of hidden nodes is set to 10. The gradient method is used for training and the resulting RMSE is 0.064. The original data and the data reconstructed from the principal components are shown in Fig. 6.26. No significant discrepancy between the original data and the reconstructed data is seen, which means that the principal components do contains the main information of the original data.

To show that a larger number of principal components will cover more information in the original data, a neural network with 20 hidden nodes has been trained on the same data. In this case, the training error is 0.044, which is smaller than the case where 10 hidden nodes are used. Theoretically, if the number of hidden nodes equals that of the inputs, the training error should equal zero.

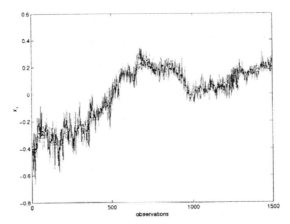

Fig. 6.26. Observations of x_1 and their approximation constructed from the principal components.

One shortcoming of using principal components analysis to reduce the input dimension is that the physical meanings of the new inputs (the principal components) are difficult to interpret. For example, the 52 inputs of the above example are the coordinates of the geometry of the blade. However, the new input with a reduced dimension is not straightforward to interpret.

6.4.5 GA-Based Input Selection

Instead of using mutual information, genetic algorithms can also be used for input selection. Application of a genetic algorithm to determine an "optimal" subset of inputs by constructing binary masks which indicate which inputs to retain and which to discard has been studied in [72]. In a conventional fuzzy system, each rule contains all the input variables in its premise. In practice, it is found that such a rule system is hard to simplify and the system performance is not satisfactory. In addition, as the number of variables increases in the rule premise, it becomes harder and harder for human beings to understand the rule.

To deal with this problem, this section employs genetic algorithms to optimize the structure of the rule base, that is, to select inputs for each fuzzy rule. Since each fuzzy rule may have different inputs, and more relevant inputs will more often be used in the fuzzy rules, this method is more flexible than mutual information based input selection, where an input is either selected or discarded.

The genetic coding for rule structure optimization is very straightforward. Suppose we are to optimize a fuzzy system consisting of N TSK fuzzy rules. The number of inputs is n. Therefore, there will be $n \times N$ genes in the chromosome, each consisting of only one binary bit, '0' or '1'. A '0' denotes that the input variable does not appear in the corresponding fuzzy rule and '1' means does. Assume there are five input variables in total and the piece of code for the j-th fuzzy rule is

$$1\ 0\ 1\ 0\ 1,$$

then, the fuzzy rule looks like:

$$R_j: \text{If } x_1 \text{ is } A_{1j},\ x_3 \text{ is } A_{3j},\ x_5 \text{ is } A_{5j}, \text{ then } y \text{ is } B_j,$$

where $A_{ij}(i = 1, 3, 5)$ are fuzzy subsets for input variable x_i in rule j and B_j is a fuzzy subset for the output y. If no input appears in the premise, the rule is removed.

To provide an illustration of the method, we consider the example in [114]. The initial fuzzy rules are generated using the direct method for data-driven fuzzy rule generation introduced in Chapter 4. Recall that there are 11 inputs in the input dimension. Let N be the number training sample, then the following error function can be used for optimizing the structure of the fuzzy system using genetic algorithms:

$$E = \frac{1}{2} \sum_{i=1}^{N} (y_i^d - y_i)^2, \tag{6.49}$$

where y_i^d is the i-th output of the system and y_i is the crisp output of the whole fuzzy system:

$$y_i = \frac{\sum_{j=1}^{M} (w_j y_j)}{\sum_{j=1}^{M} w_j}, \tag{6.50}$$

$$w_j = \prod_{i=1}^{n} g_j(A_i), \tag{6.51}$$

where M is the number of rules in the fuzzy system, $g_i(\cdot)$ is a function that defines the structure of the rule premise:

$$g_j(z) = \begin{cases} z, & \text{if } z \text{ appears in the rule premise,} \\ 1, & \text{if } z \text{ does not appear in the rule premise.} \end{cases} \tag{6.52}$$

The first try using the global performance, that is, all samples and all rules are evaluated together, was not very successful. It was not successful in the sense that the complexity of the rule system was not significantly reduced. To address this problem, local structure optimization has been considered. That is, the structure optimization is carried out inside each data patch. Suppose M_j fuzzy rules have been generated for the P data samples in patch j, then the following error function is used for evaluating the fuzzy rules in patch j:

$$E_j = \frac{1}{2} \sum_{m=1}^{P} (y_m^d - y_m)^2, \tag{6.53}$$

where P is the number of samples in data patch j, and y_m is the output of the fuzzy rules generated in data patch j:

$$y_m = \frac{\sum_{k=1}^{M_j} (w_k y_k)}{\sum_{k=1}^{M_j} w_k}, \tag{6.54}$$

$$w_j \quad = \quad \prod_{i=1}^{n} g_i(A_i). \tag{6.55}$$

Recall that maximally 2 fuzzy rules can be generated in a data patch, it is obvious that $0 \le M_j \le 2$. The number of rules for the data patch can be less than 2 because one or two of the rules may be deleted during rule similarity checking. Using this method, the complexity of the rule structure has been reduced greatly. Despite that the system has 11 inputs, the average number of variables in the premise of the structurally fuzzy rules is 3.2. Obviously, the interpretability of the fuzzy rules has been improved significantly.

6.5 Summary

This chapter presents EA-based methods for parameter and structure optimization of fuzzy systems. The main advantage of evolutionary algorithms over the traditional optimization methods is that no gradient information is needed in optimization. This makes it possible for evolutionary algorithms to optimize fuzzy models, in particular the structure of the fuzzy rules.

An evolution strategy is employed to optimize a flexible structured fuzzy controller. Scaling factors and the parameterized T-norms, and defuzzification are optimized to improve the performance of the fuzzy controller, especially for nonlinear systems.

Parameter and structure optimization of fuzzy systems using genetic algorithms are also described. It is shown that structure optimization can not only enhance the performance, but also reduce the complexity of the fuzzy systems. In fact, this kind of structure optimization is able to generate non-grid structure of fuzzy rules, which conventionally can only be obtained using recursive structure partition methods.

To handle the high dimensional systems, several methods for dimension reduction, including statistical and heuristic methods, have been introduced. An critical issue in dimension reduction for building fuzzy systems is the interpretability. Therefore, the hierarchical fuzzy systems and principal components analysis based dimension reduction are less attractive in comparison with mutual information and genetic algorithms based dimension reduction methods.

7. Knowledge Discovery by Extracting Interpretable Fuzzy Rules

7.1 Introduction

7.1.1 Data, Information and Knowledge

Data, information and knowledge are three closely related but different concepts. The following are the definitions abstracted from Webster's dictionary:

- **Data**: *1. factual information (as measurements or statistics) used as a basis for reasoning, discussion, or calculation 2. information output by a sensing device or organ that includes both useful and irrelevant or redundant information and must be processed to be meaningful 3. information in numerical form that can be digitally transmitted or processed.*
- **Information**: *1. the communication or reception of knowledge or intelligence 2. knowledge obtained from investigation, study, or instruction.*
- **Knowledge**: *1. the fact or condition of knowing something with familiarity gained through experience or association 2. acquaintance with or understanding of a science, art, or technique 3. the fact or condition of being aware of something.*

From these definitions, we understand that it is hard to clearly distinguish between data, information and knowledge. However, the following observations can be made:

- Data is a collection of facts. To get information or knowledge from data, processing may be necessary. In other words, data is a raw form of information or knowledge. Without being processed, the knowledge contained in data may not be understandable to human beings.
- Information emphasizes in a way more on communication or reception of knowledge. On the other hand, information is more than data, i.e. information can be seen as data with directions.
- Knowledge has always something to do with human understanding. "One person's data is another person's knowledge". For a machine system, knowledge can be directly transferred through data, which is not necessarily transparent to human beings. The most important point is that knowledge distinguishes from data and information in that knowledge can be reused by human or machines.

As a conclusion, it is argued that for both machine and human beings, knowledge is data or information that is understandable and can be re-used. However, what is understandable to machine may not be understandable to humans, and vice versa. For example, a neural network is able to extract knowledge from data through learning. In this process, knowledge is transferred from data to the neural network. However, the knowledge contained in the neural network is not "knowledge" to human beings until it is processed, e.g., by extracting understandable symbolic or fuzzy rules from the neural network.

In this chapter, we will discuss how knowledge understandable to humans can be extracted from data or neural networks. The extracted knowledge is represented by *interpretable* fuzzy rules. In the following, we will use the word *interpretability* to denote the conditions that fuzzy rules in a fuzzy system have to satisfy so that they can be understood by human beings. Although insufficient attention has so far been paid to the interpretability of fuzzy systems, several research results have shown that interpretability plays an important role in improving the performance of fuzzy systems and in extracting knowledge that is understandable to human beings [229, 125, 126, 13].

7.1.2 Interpretability of Fuzzy Systems and Knowledge Extraction

Knowledge extraction can be carried out by means of several different methods. In this chapter, knowledge extraction is implemented by extracting interpretable fuzzy rules from data or from neural networks. Although fuzzy systems are believed to be suitable for knowledge representation and knowledge processing because fuzzy rules are comprehensible for human beings, fuzzy rules extracted from data using evolutionary algorithms or neural networks may not be easily understandable because interpretability of the fuzzy rules can be lost.

The most important aspects of interpretability of fuzzy systems have been discussed in Chapter 1. The main aspects of interpretability of fuzzy systems include the distribution of the fuzzy partitions, including the completeness and the distinguishability, the consistency of the rule base and the complexity of the rule system. The complexity of fuzzy rules consists of the number of variables that appear in the rule premise, the number of fuzzy subset in a fuzzy partition and the number of fuzzy rules in the rule base. It is notice that the distinguishability of the fuzzy partitions and the number of fuzzy subsets within a fuzzy partition are closed related. Usually, the smaller the number of fuzzy subsets in a fuzzy partition, the more distinguishable the fuzzy partition will be.

In this chapter, major concerns will be given to the completeness and distinguishability of fuzzy partitions, the consistency of the fuzzy rules and the complexity of the rule base in addition to the approximation accuracy in fuzzy rule extraction using evolutionary algorithms and neural networks. In this way, the interpretability of the extracted fuzzy rules are preserved and knowledge in form of interpretable fuzzy rules can be extracted.

7.2 Evolutionary Interpretable Fuzzy Rule Generation

Fuzzy logic has proved to be a very powerful technique in the discipline of system control, especially when it is difficult to model the system to be controlled, or when the controlled system is of strong uncertainty and non-linearity. The fact that both successful industrial applications of fuzzy systems as well as concrete theoretical results have been reported in the last decade largely contributed to the success of fuzzy logic theory. However, fuzzy systems have also weaknesses. It has been criticized that fuzzy control rules are not capable of expressing deep knowledge [53] because they are often established on the basis of experience and intuition of human beings.

With the emergence of the techniques called soft computing or computational intelligence, fuzzy control has obtained new impetus. MPL networks, RBF networks, hybrid pi-sigma networks [115] and other neural models have been applied to the extraction of fuzzy rules from data. These methods are successful in that it is no longer necessary to determine heuristically the parameters of fuzzy rules and the fuzzy partitions of the input variables in advance.

Genetic algorithms (GAs) have also been widely employed to design fuzzy systems. Since the first attempt to vary some parameters of a fuzzy rule base using genetic algorithms [135], a huge amount of work has been done to exploit the advantages of GAs for the design of fuzzy systems. GAs are found to be very flexible because they are capable of optimizing the parameters and the rule numbers simultaneously. Furthermore, the structure of the fuzzy rules can also be optimized by GAs so that a compact fuzzy rule system can be obtained. One problem that arises in this approach is the choice of the genetic coding. If the conventional coding scheme is used, the length of the chromosome increases significantly with the number of inputs and the number of fuzzy partitions. This will no doubt harm the efficiency of the genetic search. To address this problem, chromosomes with variable length or context dependent coding [156] can be helpful.

Relatively fewer efforts have been made to date to design fuzzy systems using evolution strategies (ES). ESs are generally more suitable for the design of fuzzy systems due to their direct coding scheme and their simple way of handling constraints. In [243], evolution strategies are used to adjust the parameters of the fuzzy rules, and then genetic algorithms are utilized to optimize the structure of the fuzzy system. However, since the optimal values of the rule parameters and rule structure depend on each other, it would be better to evolve them simultaneously.

To conclude, the extraction of fuzzy rule from data can be realized with the help of neural networks or evolutionary algorithms. A common problem in data-driven rule extraction is that the interpretability of the extracted fuzzy rules can be lost. To fix this problem, not only the performance with respect to the approximation accuracy needs to be considered, but also the interpretability of the fuzzy rules should be taken into account during rule extraction.

7.2.1 Evolution Strategy for Mixed Parameter Optimization

Evolution strategies (ES) are used to optimize the parameters as well as the structure of the fuzzy rules. Evolution strategies, instead of genetic algorithms are used here due to the following two considerations. One is that the coding of ES is direct for real valued parameter optimization and consequently the length of the object vector increases just linearly with the number of variables. Since the rule structure will be evolved together with the parameters, the number of variables to be optimized increases significantly. If a binary genetic algorithm is used, the length of object vector will grow drastically. To alleviate resulting difficulties, either the range of the parameters needs to be limited, or special coding methods have to be developed. Generally speaking, evolution strategies have better performance for optimization tasks with real-valued parameters and a relatively smooth fitness landscape [6]. However, this does not imply that ES is superior to GA, or vice versa. A more detailed discussion on evolution strategies has been provided in Chapter 3.

Since both real and integer numbers are involved in the optimization, a slightly modified version of (μ,λ)-ES [4] is used here. An ES algorithm that is capable of dealing with mixed optimization can be described with the following notation:

$$(\mu, \lambda)\text{-ES} = (I, \mu, \lambda;\; m, s, \sigma;\; f, g) \tag{7.1}$$

where, I is a string of real or integer numbers representing an individual in the population, μ and λ are the numbers of the parents and offsprings respectively; σ is the parameter to control the step size, m represents the mutation operator, which is the main operator for the ES. In the ES algorithms, not only the variables, but also the step-size control parameter σ is mutated. In equation (7.1), the parameter s stands for the selection method and in this case, the parents will be selected only from the λ descendants; f is the objective function to be minimized, and g is the constraint function to which the object parameters are subject. The object parameters and the step-size control parameters are mutated in the following way:

$$\sigma_i' = \sigma_i \cdot \exp(\tau_1 \cdot N(0,1) + \tau_2 \cdot N_i(0,1)),\; i = 1, 2, ..., Q, \tag{7.2}$$

$$I_i' = I_i + \sigma_i' \cdot N_i(0,1),\; i = 1, 2, ..., Q_1, \tag{7.3}$$

$$I_i' = I_i + \lfloor \sigma_i' \cdot N_i(0,1) \rfloor,\; i = Q_1 + 1, Q_1 + 2, ..., Q, \tag{7.4}$$

where $N(0,1)$ and $N_i(0,1)$ are normally distributed random numbers with zero mean and variance of 1, Q is the total number of object parameters, Q_1

is the number of real-valued object parameters and naturally $Q - Q_1$ denotes the number of the integer object parameters. In this method, the parameters representing the fuzzy operators and membership functions are encoded with real variables, while the rule structure parameters are encoded with integer numbers. τ_1 and τ_0 are two global step size control parameters.

Several strategy parameters of ES, which have great influence on the performance of the algorithm, have to be fixed manually. These include the population size μ and λ, the global step control parameters τ_1 and τ_2, and the initial values of the step sizes σ_i. The optimization problems in the real world have normally a lot of local optima, in which a standard evolution strategy can get trapped. To acquire a best possible solution, it is desirable to improve the performance of the standard ES. In this study, a minor modification is made and, nevertheless, is proved to be effective. In practice, we find that the process of evolution stagnates when the step-size control σ converges to zero prematurely. To prevent the step-size from converging to zero, we re-initialize it with a value of, say, 1.0, when σ becomes very small. This enables the algorithm to escape from local minima on the one hand, on the other hand, it gives rise to some oscillations of the performance. Therefore, it is important to record the best individual that has been found so far. However, this best individual does not take part in the competition of selection if it does not belong to the current generation.

7.2.2 Genetic Representation of Fuzzy Systems

The genetic coding of the parameters of the fuzzy system is straightforward. Without the loss of generality, the following Gaussian membership functions are used:

$$A(x) = \exp\{-\frac{(x-c)^2}{w^2}\}. \tag{7.5}$$

Therefore, each membership function has two parameters, namely, the center c and the width w. In order to make sure that all the subsets of a fuzzy partition can distribute as freely as possible provided that the completeness condition is satisfied, the center of each fuzzy membership function can move on the universe of discourse of the corresponding variable, which is limited by the physical system. Of course, the fuzzy subsets should be ordered according to their center so that the checking of completeness can be done and that the mechanism of the rule structure optimization can work properly. The width of the membership functions is loosely limited so that it is larger than zero and naturally, not wider than the whole space. In fact, the width of the membership functions is also subject to the completeness conditions as well as the distinguishability requirements.

The coding of the rule structure is important because the size of a fuzzy system is fully specified by the rule structure. Suppose each input variable x_i has a maximal number of fuzzy subsets M_i, then the rule base has maximally

$N = M_1 \times M_2 \times \cdots \times M_n$ fuzzy rules if there are n input variables. Thus, the premise structure of the rule system can be encoded by the following matrix:

$$Struc_{premise} = \begin{bmatrix} a_{11} & a_{12} & \cdots & a_{1n} \\ a_{21} & a_{22} & \cdots & a_{2n} \\ \cdots\cdots\cdots\cdots\cdots \\ a_{N1} & a_{N2} & \cdots & a_{Nn} \end{bmatrix}, \tag{7.6}$$

where $a_{ji} \in \{0, 1, 2, ..., M_i\}, j = 1, 2, ..., N; i = 1, 2, ..., n$. The integer numbers, $1, 2, ..., M_i$ represent the corresponding fuzzy subsets in the fuzzy partition of x_i, while $a_{ji} = 0$ indicates that variable x_i does not appear in the jth rule. Note that the assumption of the maximal number of fuzzy partition will not harm the compactness of the rule system, because the redundant subsets will be discarded automatically by the algorithm during the checking of distinguishability. Similarly, the structure of the rule consequent can be encoded with a vector of positive integers:

$$Struc_{consequent} = [c_1, c_2, \cdots, c_N]^T, \tag{7.7}$$

where $c_j \in \{1, 2, ..., K\}, j = 1, 2, ..., N$, supposing that the consequent variable has maximal K fuzzy subsets. This works both for Mamdani type rules and TSK fuzzy rules whose consequent is a constant. If the TSK fuzzy rules have a real function of input variables in the consequent, the coding of the consequent structure is not necessary.

7.2.3 Multiobjective Fuzzy Systems Optimization

As we mentioned before, a fuzzy system should exhibit not only good approximation, but also good interpretability. In the fitness function, extra terms will be added to guarantee the completeness and distinguishability of the fuzzy partitions and the consistency of the fuzzy rules. The completeness of the rule structure is ensured by discarding a fuzzy membership function that is not used by any of the fuzzy rules. However, the completeness of the fuzzy partition after discarding the redundant fuzzy subset needs to be guaranteed. The completeness of the fuzzy partition of each input variable is examined using a *fuzzy similarity measure*. One advantage of using the fuzzy similarity measure is that by regulating the grade of fuzzy similarity, the degree of overlap of two subsets can be properly controlled. If the fuzzy similarity of two neighboring fuzzy subsets is zero or too small, it means that either they do not overlap or do not overlap sufficiently. On the other hand, if fuzzy similarity is too large, the two fuzzy subsets overlap too much and the distinguishability between them may be lost. To maintain both distinguishability and completeness of the fuzzy partitions, the fuzzy similarity measure of between any two neighboring membership functions is required to satisfy the following condition:

$$\underline{s} \leq S(A_i, A_{i+1}) \leq \bar{s}, \tag{7.8}$$

where A_i and A_{i+1} are two neighboring fuzzy sets, $\underline{s}$ and $\bar{s}$ are the predefined lower and upper bound of the fuzzy similarity measure. If this condition is not satisfied, a penalty will be assigned to the corresponding individual. Fig. 7.1 shows two possible penalty function that can be used to penalize incomplete or indistinguishable fuzzy rules. According to the penalty function in Fig. 7.1(a), a very large value is added to the fitness so that the individual is very unlikely to survive in selection. A better choice of the penalty function is shown in Fig. 7.1(b). Recall that for minimization problems, the smaller the fitness value, the better the performance will be.

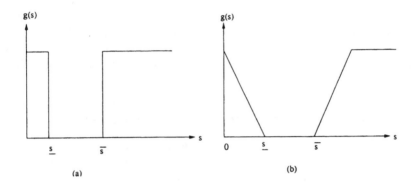

Fig. 7.1. Penalty functions for completeness and distinguishability.

The consistency index that has been suggested in Chapter 1 can not be applied directly to the evolutionary generation of fuzzy rules. To solve this problem, a degree of inconsistency of a rule base is suggested based on the consistency index. At first, a degree of *inconsistency* for the i-th rule is calculated as follows:

$$Incons(i) = \sum_{\substack{1 \le k \le N \\ k \ne i}} [1.0 - Cons(R(i), R(k))], \qquad (7.9)$$

$$i = 1, 2, ..., N$$

where, R denotes the rule base generated from data, and N is the number of rules in R. The degree of inconsistency of each rule is then summed up to indicate the degree of inconsistency of a rule base:

$$f_{Incons} = \sum_{i=1}^{N} Incons(i), \qquad (7.10)$$

which can be incorporated in the objective function of the evolutionary algorithm.

Finally, the following fitness function that combines several objectives should be used to extract interpretable fuzzy rules:

$$f = f_E + \lambda \cdot f_I + \xi \cdot f_{Incons}, \tag{7.11}$$

where f_I is the whole penalty for the completeness and distinguishability of
the fuzzy rules:

$$f_I = \sum_{i=1}^{n} \sum_{j=1}^{m_i} g(S(A_i^j, A_i^{j+1})), \tag{7.12}$$

where n is the number of input variables of the fuzzy system, m_i is the
number of fuzzy subsets in the fuzzy partition of the i-th input, and $g(s)$ is
the penalty assigned to fitness according to the similarity degree between the
j-th and $j + 1$-th subsets of the i-th input.

Note that no special measures have been taken to reduce the number of
the fuzzy rules. In practice, it is found that the evolutionary algorithm tends
to select a compact system. This implies that a standard fuzzy system with
a full grid rule structure normally has worse performance than a compact
system.

7.2.4 An Example: Fuzzy Vehicle Distance Controller

In this section, we generate a fuzzy rule system for the distance control of a
vehicle system based on collected data using the method presented above.
The diagram of the distance control system is illustrated in Fig. 7.2, where

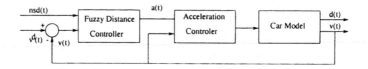

Fig. 7.2. Diagram of the distance control system.

$v(t)$ and $v_1(t)$ are the velocities of the controlled car and of the car in front
of it, $d(t)$ is the distance between the two cars, $v_r(t) = v_1(t) - v(t)$ denotes
the relative speed, and $nsd(t)$ is the normalized safety distance, which is
calculated as follows:

$$nsd(t) = \frac{d(t) - s(t)}{s(t)}, \tag{7.13}$$

where, $s(t)$ is called the safety distance. It is found that $s(t)$ is basically in
proportion to the speed $v(t)$, however, it mainly depends on the behavior of
the driver. Not only have different drivers different views on safety distance,
but the same driver also makes different decisions at different times. Our
task is to design a fuzzy distance controller that is able to produce a correct
acceleration based on the data collected from different drivers in different
driving situations. In this study, data from one driving scene are used for
training and the data from a second scene are used for test.

The following cost function is used to indicate the performance of the fuzzy controller:

$$f_E = \sum_{t=1}^{J} \sqrt{(v(t) - v^d(t))^2}, \tag{7.14}$$

where J is the total number of sampled data and v_d is the target velocity. As discussed previously, the interpretability conditions are integrated into the fitness function:

$$f = f_E + \xi \cdot f_{Incons} + \lambda \cdot f_{Incompl}, \tag{7.15}$$

where $f_{Inconsis}$ denotes the inconsistency measure and $f_{Incompl}$ is a penalty term for the completeness and distinguishability of the fuzzy system, ξ and λ are weights for the penalties. In general, once the rule system is found to be incomplete, the penalty term $f_{Incompl}$ is so large that the individual is not able to survive. That is to say, the evolutionary algorithm tolerates some degree of inconsistency or indistinguishability, but allows no incompleteness in a fuzzy system.

Based on the collected data, the meaningful range of normalized safety distance $nsd(t)$, relative speed $v_r(t)$ and the acceleration $a(t)$ are selected as [-1,5], [-10,10] and [-3,3] respectively. As mentioned before, the center of all the fuzzy membership functions is allowed to vary over the whole space of the corresponding variable. We suppose both nsd and v_r have a maximum of five fuzzy subsets and a has maximally eight fuzzy subsets. Therefore, the fuzzy system has maximally 25 fuzzy rules.

In the beginning, we generate the fuzzy rule system using 316 data collected in the first driving situation. For comparison, fuzzy rule systems are generated with and without completeness and interpretability checking. A fuzzy rule base is first generated without checking the interpretability conditions. Nevertheless, rules with the same IF-part but different THEN-part are avoided to assure the fairness of the comparison.

The inconsistency degree of the fuzzy system is 13.4 and the speed and acceleration tracking results are illustrated in Fig. 7.3. In the figure, the dotted lines denote the acceleration and speed measured in the experiment, while the solid lines describe the results produced by the fuzzy controller. It is found that the fuzzy controller has good performance on the training data, where the root-mean-square (RMS) error on acceleration is $0.105ms^{-2}$ and the RMS speed error is $0.085ms^{-1}$.

However, a fuzzy system that exhibits good performance for training data does not necessarily perform equally well on the test data. Before evaluating the fuzzy system on test data, we first have a look at the membership functions, see Fig. 7.4. We notice that some fuzzy subsets of $nsd(t)$ lack distinguishability, while the fuzzy partition of $v_r(t)$ is incomplete. This implies that over-fitting of the membership functions has occurred.

Now we evaluate the fuzzy system on the 276 test data obtained in the second driving situation. The results are presented in Fig.7.5, where the RMS

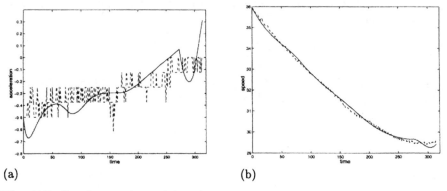

(a)   (b)

Fig. 7.3. Results on the training data without interpretability constraints: (a) acceleration; (b) speed. Solid line: experiment data; dashed line: output of the fuzzy system.

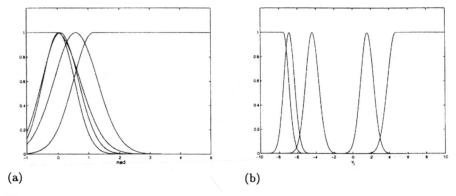

(a)   (b)

Fig. 7.4. Fuzzy partitions without interpretability constraints: (a) nsd, some of the fuzzy membership functions are indistinguishable; (b) $v_r(t)$, the fuzzy partition is incomplete.

error on the acceleration is $0.233ms^{-2}$ and on the speed is $1.044ms^{-1}$. Note that the performance of the fuzzy system on the test data has degraded seriously.

A fuzzy system is then generated with the interpretability constraints. The inconsistency degree of the fuzzy system is now reduced to 0.63. Note first that the fuzzy partition of the two input variables (see Fig. 7.6) is now complete and the distribution of the membership functions seems to be more reasonable. The acceleration and speed tracking results for the training data are demonstrated in Fig. 7.7. Compared to the fuzzy rule system generated without the interpretability constraints, the performance for the training data is quite similar. However, the performance on the test data is improved as shown in Fig. 7.8, where the RMS error on the acceleration is $0.216ms^{-2}$ and on the speed is $0.552ms^{-1}$.

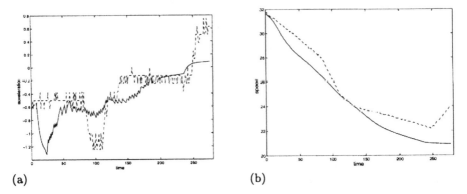

Fig. 7.5. Results on test data without interpretability constraints: (a) Acceleration; (b) speed.

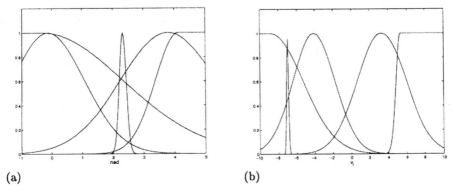

Fig. 7.6. Fuzzy partitions with interpretability constraints: (a) nsd; (b) $v_r(t)$. Both fuzzy partitions are complete.

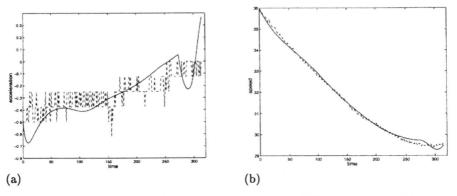

Fig. 7.7. Results on the training data with interpretability constraints: (a) acceleration; (b) speed. Solid line: experimental data; dashed line: output of the fuzzy system.

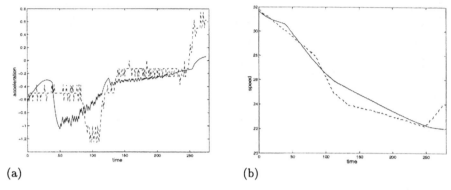

(a)   (b)

Fig. 7.8. Results on test data with interpretability constraints: (a) Acceleration; (b) speed.

From the above simulations, we see that a fuzzy system generated with interpretability constraints outperforms the one generated without checking on the test data. This means that the improvement of the interpretability may also improve the generalization performance of the fuzzy systems.

To main issues are summarized as follows. In order to extract fuzzy rules that are understandable to human beings from data using evolutionary methods, the interpretability of the fuzzy rules should be considered. The most important aspects for the interpretability of fuzzy systems include the completeness and distinguishability of the fuzzy partitions, the consistency of the fuzzy rules and the compactness of the fuzzy system. These conditions are added to the fitness function of the evolutionary algorithms. In this way, not only will the performance of the fuzzy system be optimized, but also the interpretability of the extracted fuzzy rules is improved. Besides, it has been shown that improvement of the interpretability can also improve generalization performance when less training data are available.

This approach is actually a multi-objective approach to evolutionary fuzzy rule generation and optimization. More discussions on multi-objective evolutionary optimization, including incorporation of *a priori* knowledge into evolution by means of a multi-objective approach, are provided in Chapter 7.

7.3 Interactive Co-evolution for Fuzzy Rule Extraction

7.3.1 Interactive Evolution

In evolutionary computation, the evaluation of the fitness is the essential issue in many real-world applications. One problem is that no explicit fitness function is available for evaluation of a designed system. To deal with this problem, one choice is to use interactive evolutionary computation [218]. It can be seen that interactive evolutionary computation Interactive can be done in three levels, refer to Fig. 7.9.

- Interaction in the representation level. Human interaction can directly be excised on the representation of the system. For example, in the optimization of an aerodynamic structure, an expert knows the "desired" structure on certain particular points of the structure. In this case, the human expert can directly change the representation of the blade so that good solutions will be added in the population.
- Interaction in the selection level. Usually, selection is based on a fitness evaluation and either deterministic or stochastic selection will be used. In interactive evolutionary computation, the human expert can directly select the individuals. Therefore, no fitness evaluation is necessary.
- Interaction in the fitness assignment level. An alternative to the interaction on the selection level is to assign a fitness value by the human expert to each individual. Then, a conventional deterministic or stochastic selection method can be used.

No matter which interaction scheme is used, an important issue in interactive evolution is the human-machine interface. This interface provides a means to incorporate human knowledge into evolution.

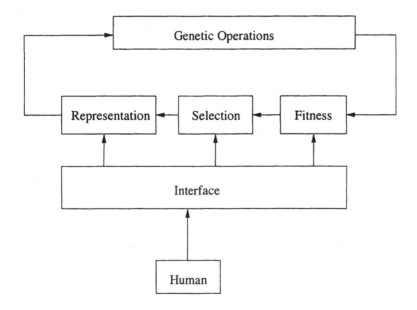

Fig. 7.9. Interactive evolution.

Interactive evolutionary computation has found a wide range of applications, such as graphics and art design [207], music generation [17] and layout design [164]. As pointed out in [218], the main difficulty in interactive evolution is how to relieve the fatigue of the involved human. One solution to this problem is the introduction of an approximate model that can model the process of the human's evaluation. In [130], a neural network has been used to model the behavior of the human to reduce human fatigue.

7.3.2 Co-evolution

In biology, co-evolution is a change in the genetic composition of one species (or sub-population) in response to a genetic change in another. In evolutionary computation, evolution usually means a number of sub-population evolve together and the fitness of one sub-population may depend on the fitness of other sub-populations.

Both competitive co-evolution and co-operative co-evolution have been studied. As the names suggest, in competitive co-evolution, the sub-population compete with each other for example, in paying games. In the cooperative co-evolution, on the other hand, all sub-populations work together to achieve a good performance for the whole system.

7.3.3 Interactive Co-evolution of Interpretable Fuzzy Systems

In extraction of interpretable fuzzy rules, the coefficient of the penalty term for completeness and distinguishability, as well the penalty term for the consistence need to be determined. Usually, trial and error methods can be used so that the a good trade-off between approximation accuracy and interpretability can be achieved.

In this section, we present an idea to determine the coefficients using another sub-population to evolve together so that an optimal value of the coefficients can be found. Whereas a smaller inconsistency is always good, it is hard to say exactly what a similarity value between two neighboring fuzzy subsets should be so that the fuzzy partition is interpretable. In this case, a decision from a human user is much better than to pre-define a value. Therefore, the interactive co-evolution should be a good choice for determination of the coefficients for extraction of interpretable fuzzy rules.

Fig. 7.10 shows framework for extraction of interpretable fuzzy rules using interactive co-evolution. Two sub-populations are involved, one for the evolution of the fuzzy rules and the other for the evolution of an optimal coefficient for the penalty term for the completeness and distinguishability of the fuzzy partitions. It can be seen that this kind of co-evolution is co-operative, because the interpretability of the extracted fuzzy rules is dependent on the evolution result of the second population.

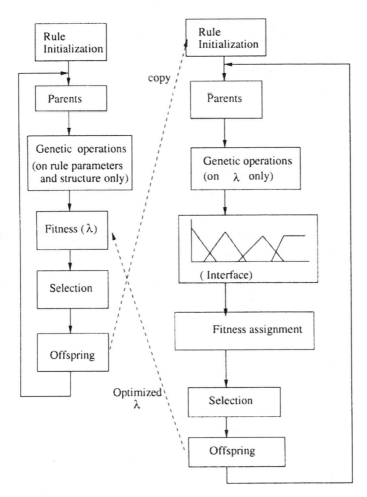

Fig. 7.10. Interactive co-evolution of interpretable fuzzy rules.

7.4 Fuzzy Rule Extraction from RBF Networks

7.4.1 Radial-Basis-Function Networks and Fuzzy Systems

Functional Equivalence. It is found in [108] that radial basis function (RBF) networks and a simplified class of fuzzy systems are functionally equivalent under some mild conditions. This functional equivalence has made it possible to combine the features of these two models systematically and has initiated a new model called neurofuzzy system [109] . However, we believe that such neurofuzzy models will be more attractive if they preserve the main features of both neural networks and fuzzy inference systems. That is to say, they should have good learning capability, strong representational power as well as good interpretability (transparency). Among these features, learning capability and representational power are always guaranteed since they are the inherent features of neurofuzzy systems. On the contrary, the interpretability of neurofuzzy systems is not inherent and relatively little attention has been paid to it. In fact, interpretability of neurofuzzy systems is usually impaired due to excessive endeavors for the improvement of approximation performance of neurofuzzy systems.

Consider an RBF network with the output normalized:

$$y = \frac{\sum_{j=1}^{N} f_j \phi_j(\| \mathbf{x} - \boldsymbol{\mu}_j \| / \sigma_j)}{\sum_{j=1}^{N} \phi_j(\| \mathbf{x} - \boldsymbol{\mu}_j \| / \sigma_j)} \tag{7.16}$$

This normalized model of RBF networks has very interesting properties, one of which is its mathematical equivalence to the TSK fuzzy models [108]. The TSK fuzzy model has the following form of fuzzy rules:

$$R_j : \text{If } x_1 \text{ is } A_{1j} \text{ and } x_2 \text{ is } A_{2j} \text{ and } ... \text{ and } x_M \text{ is } A_{Mj},$$
$$\text{then } y = g_j(x_1, x_2, ..., x_M)$$

where $g_j(\cdot)$ is a crisp function of x_i. The overall output of the fuzzy model can be obtained by:

$$y = \frac{\sum_{j=1}^{N} (g_j w_j)}{\sum_{j=1}^{N} w_j}, \tag{7.17}$$

where w_j is the truth value of the rule. A schematic description of a fuzzy system with two input variables is given in Fig. 7.11, where each variable has two fuzzy subsets and there are four fuzzy rules in total.

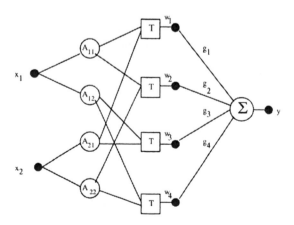

Fig. 7.11. The TSK fuzzy model with two inputs and each input has two fuzzy subsets.

It is noticed that the normalized RBF network model and the TSK fuzzy model are mathematically equivalent provided that multiplication is used for the t-norm in fuzzy systems and both systems use Gaussian basis functions (refer also to Fig. 7.12). We will not re-state the restrictions proposed in [108], however, we will show that although these restrictions do result in the mathematical equivalence between RBF networks and fuzzy systems, they do not guarantee the equivalence of the two models in terms of the physical meaning. In fact, when we talk about extracting fuzzy rules from neural networks, there should be some differences between the RBF networks and TSK fuzzy systems.

Interpretability: the Difference between RBFN and TSK Model. The main difference between radial-basis-function networks and fuzzy systems is the interpretability. Generally speaking, neural networks are considered to be black-boxes and therefore no interpretability conditions are imposed on conventional neural systems. Therefore, RBF networks and TSK fuzzy models have essential differences despite their mathematical equivalence. These differences are embodied in terms of constraints on parameters of interpretable fuzzy systems. In order to convert an RBF network into an interpretable fuzzy system, in other words, to extract interpretable fuzzy rules from RBF networks, the following conditions should be satisfied:

1. The basis functions of the RBF network are Gaussian functions.
2. The output of the RBF network is normalized.
3. The basis functions within each receptive field unit of the RBF network are allowed to have different variances.
4. Certain numbers of basis functions for the same input variable but within different receptive field units should share a mutual center and a mutual variance.

Conditions 3 and 4 are essential for good interpretability of the extracted fuzzy system. Without condition 3, condition 4 can not be fulfilled. As the most important condition, condition 4 requires that some weights in the RBF network should share, which in effect limits the number of basis functions for each variable, thus to limit the number of fuzzy subsets within the fuzzy partitions, when the RBF network is converted into a a fuzzy system. It is obvious that a smaller number of fuzzy subsets in a fuzzy partition usually leads to a better distinguishable fuzzy partition. Thus, the interpretability of the fuzzy system is improved, provided the fuzzy partition is complete.

For the sake of simplicity, we use 'weight' to refer to both parameters of basis functions (centers and variances) and the output weights of RBF networks in the following text.

Weight sharing is a necessary condition for the good distinguishability of the fuzzy partitions. Without this condition, each rule in a fuzzy system will use a different membership function for each variable in the rule premise. This has two consequences:

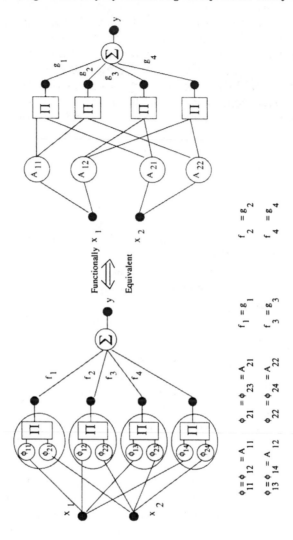

Fig. 7.12. Functional equivalence between RBF networks and fuzzy systems.

1. The number of membership functions within the fuzzy partition of the input variables will be very large;
2. The fuzzy subsets within the fuzzy partition are hardly distinguishable.

Both consequences degrade the interpretability of fuzzy systems. In an RBFN, the number of the basis functions usually equals the number of the hidden nodes, if no weights share. If the RBF network is to be treated as a fuzzy system, the number of the basis functions should be reduced in a way that some of the basis functions share the same value. This can be easily found if we take a look at the RBF network converted from a interpretable fuzzy system. For example, in Fig.7.12, $\phi_{11} = \phi_{12}$, $\phi_{21} = \phi_{23}$ and so on.

Unfortunately, it is unreasonable to define weight sharing conditions explicitly because we cannot predefine the structure of the extracted fuzzy system. Therefore, we do not know beforehand which weights should share when an RBFN is converted to a interpretable fuzzy system. Additionally, the completeness of the fuzzy partitions should be considered together with the weight sharing condition in the course of rule extraction. These problems will be treated in detail in the next section.

Note that the consistency condition is not considered here, since it is assumed that the knowledge acquired by the RBF network is consistent. Nevertheless, measures can be taken to prevent the rule extraction algorithm from generating seriously inconsistent rules, namely, rules with the same premise but different consequent parts.

7.4.2 Fuzzy Rule Extraction by Regularization

As we have discussed in the last section, to extract interpretable fuzzy rules from an RBF network, some weights should share. That is to say, there exist some inherent constraints among the parameters of a fuzzy system. As shown in Fig. 7.12, the effective number of free parameters in a fuzzy system is smaller than that in an RBF network, although they are mathematically equivalent. Thus, extracting interpretable fuzzy rules from RBF networks can be regarded in a way as reducing the effective number of free parameters in the neural network. To achieve this, neural network regularization methods can be employed.

The original goal of regularization was to improve the generalization ability of neural networks. One basic idea of regularization is to reduce the effective number of free parameters in the neural network. Weight sharing [153] is one of the methods inspired by this idea, which tries to drive different but similar weights (defined by the Euclidean distance) to one common value. Therefore, it is necessary to specify, which weights should be identical before the weight sharing algorithm can be applied. Thus, the first step towards rule extraction from RBF networks is to determine which weights, including the parameters of basis functions and the output weights, should share.

Specification of Weights to Share. Since the rule structure of the fuzzy system is unknown, we are not aware in advance, which weights should be identical. However, it is straightforward to imagine that the distance between the weights that will have a common value should be small. Therefore, we have to identify similar weights using a distance measure. Currently, several distance measures (or similarity measures) are available. Among them, the Euclidean distance is very simple and has been widely used. For two membership functions (basis functions) $\phi_i(\mu_i, \sigma_i)$ and $\phi_j(\mu_j, \sigma_j)$, where μ_i, μ_j are centers and σ_i, σ_j are the variances, the Euclidean distance between them is defined by:

$$d(\phi_i, \phi_j) = \sqrt{(\mu_i - \mu_j)^2 + (\sigma_i - \sigma_j)^2}. \tag{7.18}$$

For the output weights, which are fuzzy singletons, there is only one element. The procedure to find out the weights to share is described in the following:

1. Order the basis functions (ϕ_{ij}) increasingly according to their center value. Let U_{ik} denote the k-th set for x_i whose elements (basis functions) have a distance less than d_i and should share, where d_i is a prescribed threshold. Put ϕ_{ij} in U_{ik}, let $j, k = 1$, $\phi_i^0 = \phi_{ij}$.
2. If the distance $d(\phi_i^0, \phi_{ij+1}) < d_i$, where d_i is a predefined threshold, put ϕ_{ij+1} in U_{ik}; else $k = k + 1$, put ϕ_{ij+1} in U_{ik} and let $\phi_i^0 = \phi_{ij+1}$.
3. $j = j + 1$, if $j \leq M$, go to step 2; else stop.

Refer to Fig. 7.13 for an illustration of the specification of weights to share.

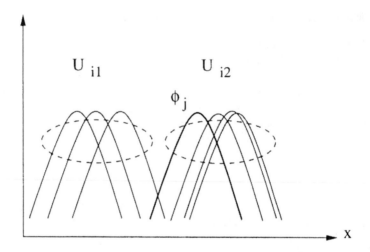

Fig. 7.13. Clustering of the similar weights.

Since the basis functions in U_{ik} will share a mutual center and a mutual variance, the prescribed distance threshold d_i is very important. It determines not only the distinguishability, but also the completeness of the fuzzy partition. Suppose $\hat{\mu}_{ik}$ and $\hat{\sigma}_{ik}$ are the average center and variance of the basis functions in U_{ik}, then the fuzzy partition composed by $\hat{\mu}_{ik}$ and $\hat{\sigma}_{ik}$ should satisfy the completeness and distinguishability condition.

In practice, we find that the performance of the extracted fuzzy system is not satisfactory if we simply select $\hat{\mu}_{ik}$ and $\hat{\sigma}_{ik}$ as the values to share by the basis functions in U_{ik}. In other words, it is imperative to adapt the parameters to share. In the following subsection, we will introduce an adaptive weight sharing method to improve the performance of the extracted fuzzy system.

Adaptive Weight Sharing. We do not directly require that the weights in the same set U_{ik} should be identical. Alternatively, we realize weight sharing by regularizing the RBF network. In order to describe the regularization algorithm clearly, we assume that the RBF network has the following form:

$$y = \frac{\sum_{j=1}^{N}[w_j \prod_{i=1}^{m_j} \exp(-\frac{(x_i-\mu_{ij})^2}{\sigma_{ij}^2})]}{\sum_{j=1}^{N} \prod_{i=1}^{m_j} \exp(-\frac{(x_i-\mu_{ij})^2}{\sigma_{ij}^2})}, \tag{7.19}$$

where, N is the number of the receptive field units (the number of fuzzy rules), m_j is the number of basis functions within the j-th receptive field unit (the number of variables appearing in the premise of the j-th rule), μ_{ij} and σ_{ij} are the center and variance of the basis function (fuzzy membership function), and w_j is the output weight of the j-th receptive field unit.

It is noticed that we do not use a grid rule structure for the fuzzy system to provide more flexibility. In addition, we intend to optimize both the parameters and the structure of the fuzzy system with the help of neural network learning.

Regularization of neural networks is realized by adding an extra term to the conventional cost function:

$$J = E + \lambda \cdot \Omega, \tag{7.20}$$

where, E is the conventional cost function, λ is the regularization coefficient $(0 \leq \lambda < 1)$, and Ω is the regularization term for weight sharing. The cost function E is expressed as:

$$E = \frac{1}{2}(y - y^d)^2, \tag{7.21}$$

where, y is the output of the neural network and y^d is the desired value. We first suppose that TSK fuzzy rules are expected to be extracted, that is to say, the output weights of the RBF network are not regularized. In this case, the regularization term Ω has the following form:

$$\Omega = \frac{1}{2}\sum_{i}\sum_{k}\sum_{\phi_{ij} \in U_{ik}} (\mu_{ij} - \bar{\mu}_{ik})^2 + \frac{1}{2}\sum_{i}\sum_{k}\sum_{\phi_{ij} \in U_{ik}} (\sigma_{ij} - \bar{\sigma}_{ik})^2 \tag{7.22}$$

where $\bar{\mu}_{ik}$ and $\bar{\sigma}_{ik}$ are the center and variance to share by the basis functions ϕ_{ij} in set U_{ik}. Empirically, the average center $\hat{\mu}_{ik}$ and variance $\hat{\sigma}_{ik}$ of set U_{ik} are used as the initial values for $\bar{\mu}_{ik}$ and $\bar{\sigma}_{ik}$. Thereafter, $\bar{\mu}_{ik}$ and $\bar{\sigma}_{ik}$ are not directly dependent on the value of μ_{ik} and σ_{ik}. The gradient of J with respect to the center and the variance can be calculated as follows:

$$\frac{\partial J}{\partial \mu_{ij}} \mid_{\phi_{ij} \in U_{ik}} = \frac{\partial E}{\partial \mu_{ij}} + \lambda(\mu_{ij} - \bar{\mu}_{ik}), \tag{7.23}$$

$$\frac{\partial J}{\partial \sigma_{ij}} \mid_{\phi_{ij} \in U_{ik}} = \frac{\partial E}{\partial \sigma_{ij}} + \lambda(\sigma_{ij} - \bar{\sigma}_{ik}), \tag{7.24}$$

$$\frac{\partial J}{\partial \bar{\mu}_{ik}} = -\lambda \sum_{\phi_{ij} \in U_{ik}} (\mu_{ij} - \bar{\mu}_{ik}), \tag{7.25}$$

$$\frac{\partial J}{\partial \bar{\sigma}_{ik}} = -\lambda \sum_{\phi_{ij} \in U_{ik}} (\sigma_{ij} - \bar{\sigma}_{ik}). \tag{7.26}$$

The corresponding gradients of the error E are:

$$\frac{\partial E}{\partial \mu_{ij}} =$$

$$(y - y^t)(w_j - y)(x_i - \mu_{ij}) \prod_{k=1}^{m_j} \left[\exp \left(-\frac{(x_k - \mu_{ij})^2}{\sigma_{ij}^2} \right) \right] / \sigma_{ij}^2, \tag{7.27}$$

$$\frac{\partial E}{\partial \sigma_{ij}} =$$

$$(y - y^t)(w_j - y)(x_i - \mu_{ij})^2 \prod_{k=1}^{m_j} \left[\exp \left(-\frac{(x_k - \mu_{ij})^2}{\sigma_{ij}^2} \right) \right] / \sigma_{ij}^3. \tag{7.28}$$

It should be pointed out that the specification algorithm introduced in the last section must be applied in each iteration of the neural network learning to improve the performance of the extracted fuzzy system. Recall that the weights of the RBF network are determined not only by the regularization term, but also by the conventional error term. That is to say, it is possible that a basis function ϕ_{ij} that is originally put in U_{ik} will be moved to another set after an iteration of learning, refer to Fig. 7.14 for an illustration. By specifying the shared weights during network learning, a more optimal rule structure will be obtained. It is also necessary to check the completeness condition for the fuzzy partitioning constructed by $(\bar{\mu}_{ik}, \bar{\sigma}_{ik})$ during learning. If incompleteness is observed during regularization, it will be necessary to adjust the distance threshold d_i, see step 1 in specification of weighs to share.

If Mamdani fuzzy rules are to be extracted, the specification of weights to share should also be carried out for the output weights (w_j) and an extra term should be added to Ω. Then, a regularized learning algorithm for the output weights can also be derived. The resulting weights can be explained as the center of the membership functions for the output variable.

A flowchart of the rule extraction process is illustrated in Fig. 7.15.

The extraction of interpretable fuzzy rules from the RBF network establishes a bi-directional bridge between RBFNs and fuzzy systems. Based on heuristic knowledge and experimental data, a fuzzy system can be constructed and its structure can be optimized using evolutionary algorithms. Then the fuzzy system can be converted to an RBFN based on the mathematical equivalence between the RBFNs and fuzzy systems. The parameters

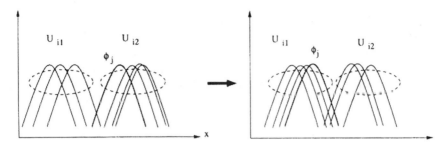

Fig. 7.14. During learning, membership ϕ_j (denoted by thick line) is changed from cluster U_{i1} to U_{i2}.

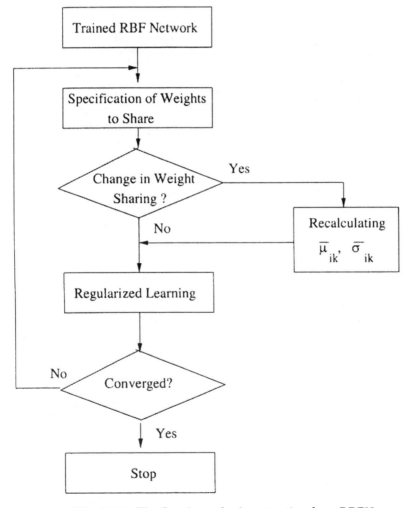

Fig. 7.15. The flowchart of rule extraction from RBFN.

of the RBFN can further be refined using the neural networks learning algorithms. In this process, the interpretability of the original fuzzy system is most probably lost. With the help of the rule extraction algorithm presented above, the RBFN can again be converted to interpretable fuzzy rules, thus knowledge can be extracted from the neural network. Since data as well as the *a priori* knowledge have been used during this process, the extracted knowledge is no double of a higher quality and deeper than the *a priori* knowledge. Refer to Fig. 7.16 for an illustration of this process of knowledge embedding and knowledge extraction.

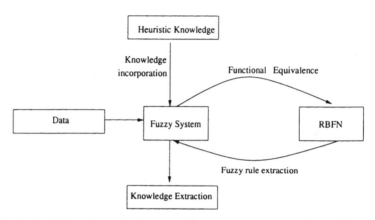

Fig. 7.16. A bi-directional bridge between fuzzy systems and neural networks, between knowledge knowledge incorporation and knowledge extraction.

7.4.3 Application Examples

The Mackey-Glass System. In this subsection, simulation studies on a prediction of the Mackey-Glass time series are carried out to show the feasibility of the proposed method. The Mackey-Glass time series is described by:

$$\dot{x} = \frac{ax(t - \tau)}{1 + x^b(t - \tau)} - cx(t), \tag{7.29}$$

where $\tau = 30$, $a = 0.2$, $b = 10$, and $c = 0.1$. One thousand data samples are used in the simulation, 500 samples for training and the other 500 samples for test. The goal is to predict $x(t)$ using $x(t-1)$, $x(t-2)$ and $x(t-3)$. That is to say, the system has three inputs and one output.

An RBF neural network with 6 hidden nodes is converted from a fuzzy system that is generated using the training data [125]. After training of the RBF network, the mean absolute errors on the training and test data are about 0.008. The basis functions of the $x(t-2)$ and $x(t-1)$ are shown in

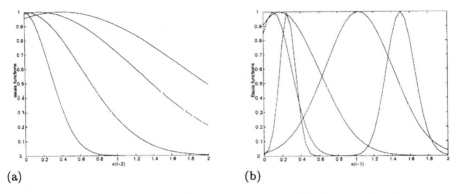

Fig. 7.17. Basis functions of the RBF network. (a) $x(t-2)$, (b) $x(t-1)$.

Fig. 7.17. $x(t-3)$ has only one basis function, and therefore, it is not shown in the figure.

The weight specification process is then applied on the basis functions of $x(t-2)$ and $x(t-1)$. As a result, the basis functions of $x(t-2)$ are classified into two groups and those of $x(t-1)$ are divided into three groups. Using the adaptive weight sharing algorithm, five fuzzy rules are obtained. The mean absolute errors for training and test are both about 0.016, which are larger than those of the RBF network. The new basis functions of $x(t-3)$, $x(t-2)$ and $x(t-1)$, which are now called membership functions, are shown in Fig. 7.18.

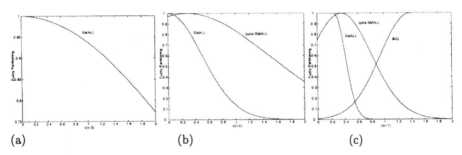

Fig. 7.18. Membership functions of the fuzzy system. (a) $x(t-3)$, (b) $x(t-2)$, and (c) $x(t-1)$

According to the distribution of the membership functions, a linguistic term is assigned to each membership function of the inputs, refer to Fig. 7.18. For example, for $x(t-2)$, "SMALL" can be assigned to its membership function $(0.01, 0.7)$ and "quite SMALL" can be assigned to the membership function $(0.29, 1.68)$. In this way, the following interpretable fuzzy rules for the Mackey-Glass system can be extracted:

- If $x(t-2)$ is SMALL and $x(t-1)$ is BIG, then $x(t)$ is BIG;
- If $x(t-1)$ is quite SMALL, then $x(t)$ is SMALL;
- If $x(t-2)$ is quite SMALL and $x(t-1)$ is SMALL, then $x(t)$ is SMALL;
- If $x(t-1)$ is BIG, then $x(t)$ is BIG;
- If $x(t-3)$ is SMALL and $x(t-2)$ is SMALL, then $x(t)$ is BIG.

It is interesting to notice that the extracted fuzzy rules mainly cover the extreme points of the series. This coincides with the analytical results drawn from some function approximation problems in [142].

The Lorenz System. The goal of the simulation is to show how an interpretable fuzzy system can be extracted from a trained RBF network. Through such a process, we are able to discover unknown dependencies between the input and output of the system and thus to acquire understandable knowledge about it.

The Lorenz system studied in this section is described by the following differential equations [161]:

$$\frac{dx}{dt} = -y^2 - z^2 - a(x - F) \tag{7.30}$$

$$\frac{dy}{dt} = xy - bxz - y + G \tag{7.31}$$

$$\frac{dz}{dt} = bxy + xz - z, \tag{7.32}$$

where $a = 0.25$, $b = 4.0$, $F = 8.0$ and $G = 1.0$. In our simulation, we predict $x(t)$ from $x(t-1)$, $y(t-1)$ and $z(t-1)$. Using the fourth order Runge-Kutta method with a step length of 0.05, 2000 data pairs are generated, where 1000 pairs of data are used for training and the other 1000 for test.

An RBF network with 5 hidden nodes is converted from a fuzzy system [127]. In the network, $x(t)$ has 5 basis functions, both $x(t-1)$ and $y(t-1)$ have 3 basis functions, which are shown in Fig. 7.19. It will be difficult to assign a linguistic term to the basis functions if they are directly treated as fuzzy membership functions, because some of the basis functions are very similar and difficult to distinguish.

The proposed algorithm for extracting fuzzy rules is then employed. The algorithm consists of two parts: specification of shared weights and network training with regularization. Although the performance of the extracted fuzzy rules is a little worse than that of the RBF network (Fig. 7.20), interpretability of the extracted fuzzy model is much better: the fuzzy partitions are both complete and well distinguishable (see Fig.7.21) and the number of fuzzy rules is also reduced.

With these interpretable fuzzy rules at hand, knowledge about the underlying system can be acquired. In the fuzzy model, $x(t-1)$ has 4 fuzzy subsets, whereas $y(t-1)$ has 2 subsets and $z(t-1)$ has only one subset. It is straightforward to see that in this system, $x(t)$ depends much more on $x(t-1)$ than on $y(t-1)$ or $z(t-1)$. According to the distribution of the membership

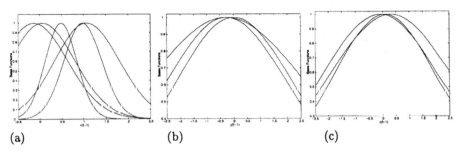

(a)   (b)   (c)

Fig. 7.19. The basis functions of the trained RBF network: (a) $x(t-1)$, (b) $y(t-1)$ and (c) $z(t-1)$.

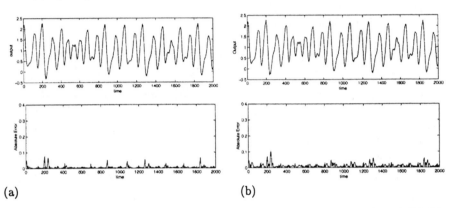

(a)   (b)

Fig. 7.20. Output and approximation error: (a) The RBF network and (b) The extracted fuzzy model.

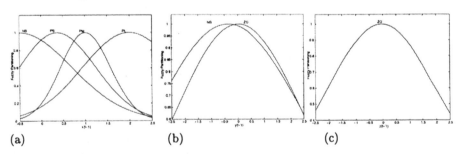

(a)   (b)   (c)

Fig. 7.21. The membership functions of the extracted fuzzy model: (a) $x(t-1)$, (b) $y(t-1)$ and (c) $z(t-1)$.

functions, a proper linguistic term can be assigned to each fuzzy set, refer to Fig. 7.21. For input $x(t-1)$, "**Negative Small (NS)**" can be assigned to the membership function $(-0.49, 1.69)$, "**Positive Small (PS)**" to $(0.37, 1.24)$, " **Positive Middle (PM)**" to $(1.01, 0.80)$ and "**Positive Large (PL)**" to $(2.01, 1.49)$. Similarly, for $y(t-1)$, "**Negative Small (NS)**" is assigned to $(-0.35, 4.11)$, "**Zero (ZO)**" to $(0.03, 3.53)$; for $z(t-1)$, "**Zero (ZO)**" is assigned to $(-0.04, 2.79)$. In this way, some intelligible knowledge about the Lorenz system in terms of interpretable fuzzy rules is acquired.

- If $x(t-1)$ is **Positive Small**, then $x(t)$ is **Negative Small**
- If $x(t-1)$ is **Positive Large** and $y(t-1)$ is **Zero** and $z(t-1)$ is **Zero**, then $x(t)$ is **Positive Large**
- If $x(t-1)$ is **Negative Small** and $z(t-1)$ is **Zero**, then $x(t)$ is **Negative Small**
- If $x(t-1)$ is **Positive Middle** and $y(t-1)$ is **Negative Small**, then $x(t)$ is **Positive Middle**

We mentioned that the approximation accuracy of the extracted fuzzy system is a little worse than that of the RBF neural network. This implies that better interpretability may lead to a lower approximation accuracy on the training data, especially when the fuzzy partitions of the fuzzy system are required to be well distinguishable. On the other hand, improvement of interpretability can in some cases improve the generalization performance.

Process Modeling. The data used in the following simulation are generated to simulate an industrial process. We use this example because it is a high-dimensional system with deliberately added biased noises. In this simulated system, there are 11 inputs and one output with 20,000 data for training and 80,000 data for test.

An RBF network with 27 hidden nodes is obtained using the training data. The RMS errors on training and test data are 0.189 and 0.207. Although the performance is satisfying (with reference to the theoretically possible minimal error), the RBF model is hard to understand when we have a look at the basis functions, particularly 8 of the 11 inputs, see Figures 7.22, 7.23, 7.24 and 7.25.

To extract interpretable fuzzy rules from the RBF network, the proposed algorithm is employed. The RMS errors of the fuzzy system on the training and test data are 0.191 and 0.213, respectively, which have slightly increased as expected. What is very encouraging is that the number of fuzzy subsets in the fuzzy partitions are significantly reduced and the distinguishability is greatly improved, see Figures 7.26, 7.27, 7.28 and 7.29. With these well distinguishable fuzzy partitions, it is possible to associate a linguistic term to each fuzzy subset and thus interpretable fuzzy rules can be obtained. This demonstrates that the proposed algorithm is effective also for high-dimensional systems.

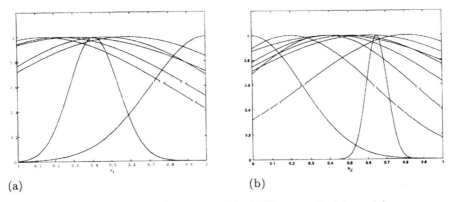

(a)   (b)

Fig. 7.22. The basis functions of the RBF network: (a) x_1, (b) x_2.

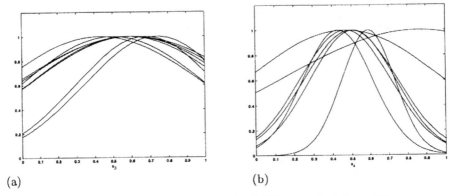

(a)   (b)

Fig. 7.23. The basis functions of the RBF network: (a) x_3, (b) x_4.

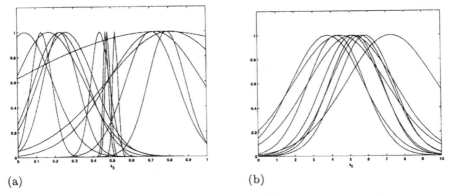

(a)   (b)

Fig. 7.24. The basis functions of the RBF network: (a) x_5, (b) x_6.

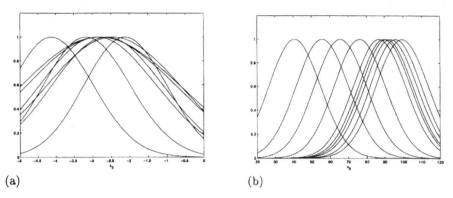

Fig. 7.25. The basis functions of the RBF network: (a) x_8, (b) x_9.

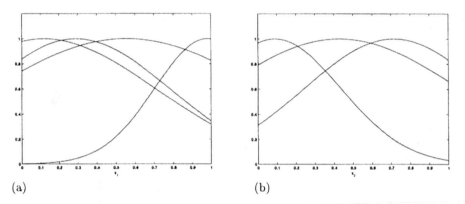

Fig. 7.26. The membership functions of the fuzzy system: (a) x_1, (b) x_2.

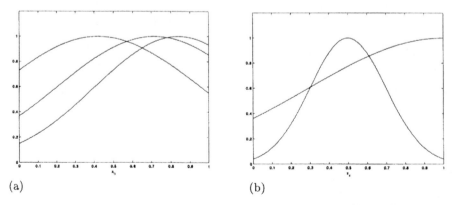

Fig. 7.27. The membership functions of the fuzzy system: (a) x_3, (b) x_4.

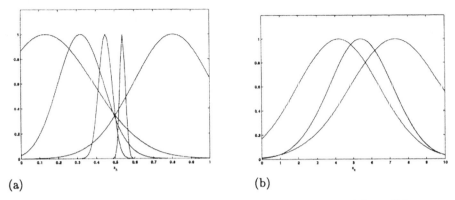

Fig. 7.28. The membership functions of the fuzzy system: (a) x_5, (b) x_6.

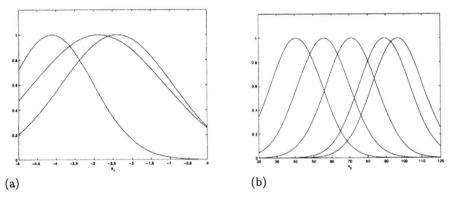

Fig. 7.29. The membership functions of the fuzzy system: (a) x_8, (b) x_9.

7.5 Summary

This chapter describes how knowledge can be acquired from data by extracting interpretable fuzzy rules using evolutionary algorithms and neural networks. To ensure that the extracted fuzzy rules are understandable to human beings, the concept of interpretability of fuzzy systems incorporated in the fuzzy rule generation. If interpretable fuzzy rules are to be generated from data using evolutionary algorithms, the interpretability conditions are integrated in the fitness function. A regularized learning method is developed to extract interpretable fuzzy rules from RBF networks. With these two methods, knowledge can be extracted from data in the form of understandable fuzzy rules.

In extracting interpretable fuzzy rules from data, one may have to reach a tradeoff between the performance and the interpretability. In general, the interpretability is improved at the sacrifice of the performance, which can be adjusted by the parameters in the fitness function or in the cost function. On the other hand, it is shown that in some cases, e.g., when training data are sparse, the generalization performance can also be improved by improving the interpretability.

8. Fuzzy Knowledge Incorporation into Neural Networks

8.1 Data and *A Priori* Knowledge

Conventional neural network learning algorithms use data as the only source of knowledge, no matter whether the supervised, the unsupervised or the reinforcement learning is employed. This is due to the fact that the conventional learning algorithms are derived from statistics or probability theory, both of which are strongly dependent on data samples.

Unfortunately, collecting a sufficient number of data is not always feasible. This difficulty may be caused by the following different factors:

- Collection of data is computationally expensive. One good example is the Computational Fluid Dynamics (CFD) simulation. One two-dimensional (2-D) CFD simulation using the Navier-Stokes equations will take about 30 minutes of CPU time on a high-performance computer. Much worse, a 3-D CFD simulation will take as many as 10 CPU hours, which makes data collection tremendously difficult.
- Collection of data needs experiments that are prohibitive. In many industrial processes, it is very difficult to carry out a control experiment to collect sensible data. Sometimes one has to interrupt the normal running of the process to get data, which is not allowed in many cases. Another example is data collection in robotics. One has to let the robot perform a certain task to collect data, which is not only time-consuming, but is also possibly hazardous to the robot.

On the other hand, large amount of *a priori* knowledge may be available about a given system. Generally, *a priori* knowledge can be divided into three classes:

- *Heuristics*: Heuristics is a kind of common sense. For example, it is intuitive to have the following knowledge:

    If the speed of a car is high, then brake forcefully to stop it.
    If the temperature is higher, then the pressure inside a tire is higher.

- *Expert knowledge*. Expert knowledge is accumulated by a human expert in practice. For example, a control expert knows that *if the rise-time of the system response is too large, then the proportional part of a PID controller should be increased* or *if the system response oscillates strongly, then the differential part of the PID controller should be reduced.*

- *Hints.* Hints can be some background knowledge about the system. One type of often used hints is the invariance property in image processing. In approximating a target function, knowledge such as *the function is monotonically increasing or the function is symmetric* can be seen as a hint. In predicting running water consumption, one may have such a hint as *in early morning, the water consumption will be high and in mid-night, it will be low.* In controlling of a system, it is easy to get the hints as *the output increases/decreases when the control input increases.*

- *Machine knowledge.* In some cases, *a priori* knowledge can also be transferred from other learning and/or evolution processes. For example, knowledge from a surrogate (approximate model) such as a neural network that is trained using history data can also be used in evolution and learning. This kind of knowledge may not be understandable to human beings, nevertheless, it plays an important role in knowledge transfer in evolutionary and learning systems.

In this chapter, we mainly discuss the utilization of human knowledge that is not directly in the form of data. A number of methods can be used to incorporate *a priori* knowledge into neural networks. Some widely used methods are:

- *Network initialization.* A priori knowledge may be used to initialize the neural network model, including the parameters, the structure and the activation functions. For example, one can construct a fuzzy rule system using expert knowledge. Based on the mathematical equivalence between a class of fuzzy systems and RBF neural networks, the fuzzy system can be used as the initial model for the neural network, both for the structure and parameters.

- *Regularized learning.* An extra term can be added to the conventional error function. This term penalizes the system if it learns something that does not agree with the *a prior* knowledge.

- *Multi-task learning.* It has been noticed that in the learning process of humans, it is helpful if several related tasks are learned simultaneously. This also applies to artificial learning systems [225]. Thus, *a priori* knowledge, if properly represented, can be used in the form of a related task [123]. Multi-task learning is also known as catalyzed learning and a related task is also called catalyst [29].

The methods discussed above can applied also to evolutionary systems. In [126], heuristics is represented by means of fuzzy rules and an extra term is added to the fitness function to check the consistency between the fuzzy rules obtained from heuristics and the rules generated from data. A related task (catalyst) in learning systems has been regarded as a problem that is belong to a *problem class* in evolution [99]. However, there are additional ways to

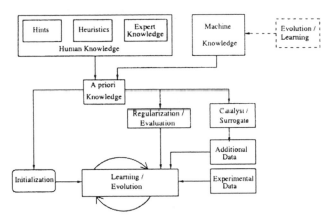

Fig. 8.1. Data and *a priori* knowledge in evolutionary and learning systems.

incorporate knowledge into evolutionary systems. For example, interactive evolution [75], or evolution using approximate models [119, 118]. If one takes a closer look at these methods, one can find that the interactive evolution is comparable to the regularized learning (both affect the learning or evolution process by modifying the cost function or the fitness function). Meanwhile evolution with approximate models (surrogates) is corresponding to learning with catalysts (additional data are generated from *a priori* knowledge). From the above discussions, the role of data and *a priori* knowledge in learning and evolutionary systems can be described by Fig. 8.1.

In this chapter, the incorporation of *a priori* knowledge into data-driven learning systems is discussed. The knowledge is represented with a number of fuzzy rules extracted from human heuristics.

8.2 Knowledge Incorporation in Neural Networks for Control

8.2.1 Adaptive Inverse Neural Control

Neural networks have widely been used in control of complex systems [96]. Generally, the function of neural networks can be divided into the following categories:

- Direct or adaptive inverse control. For systems control, if the inverse dynamics of the controlled plant is known, then it is easy to realize any control trajectories. However, most plants in real-world applications are of strong nonlinearity and high uncertainties, therefore, it is very effective to use neural networks to learning the inverse dynamics of a control plant. If the neural network is only trained off-line, it is known as direct inverse control. In contrast, if the neural network is adapted on-line, it is then called adaptive inverse control, refer to Fig. 8.2.

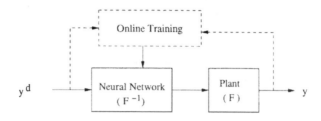

Fig. 8.2. Direct and adaptive inverse neural control. The dashed lines denotes the components for adaptive inverse control.

- Combination with conventional model-based control. Most modern control methods are model-based, where neural networks can be very helpful. For example, neural networks have been applied to model reference adaptive control [173], internal model control [97] and nonlinear feedback control [129], to name a few.
- Dynamic programming and optimal control. Due to the optimization capability of neural networks, they have also been used in dynamic programming and optimal control [102].
- Parameter estimation and systems identification.
- Reinforcement control. If the desired control is not directly available, reinforcement control is very effective [9] .

8.2.2 Knowledge Incorporation in Adaptive Neural Control

Consider the following system to be controlled:

$$x_i(k+1) = f_i(x_1(k), ..., x_1(k - P_1^i + 1), ..., x_n(k), ..., x_n(k - P_n^i + 1),$$
$$u_1(k), ..., u_1(k - Q_1^i), ..., u_m(k), ..., u_m(k - Q_m^i)), \qquad (8.1)$$

where $x_i(i = 1, ..., n)$ are system states, and $u_j(j = 1, ..., m)$ are control variables. Function $f_i(\cdot)$ is a linear or nonlinear function of the system states and controls, P_j^i and Q_j^i are non-negative structure parameters of the system. To simplify the discussion, no structure identification of the system will be involved. Therefore, it is assumed that the structure parameters are known.

Suppose the system in equation (8.1) is invertible. Thus, for each unknown system $f_i(\cdot)$, there exists a function $g_i(\cdot)$, such that

$$u_i(k) = g_i(x_1(k+1), x_1(k), ..., x_1(k - P_1^i + 1), ..., x_n(k+1), x_n(k), ...,$$
$$x_n(k - P_n^i + 1), u_1(k - 1), ..., u_1(k - Q_1^i), ..., u_m(k - 1), ..., u_m(k - Q_m^i)),$$
$$(8.2)$$

where the system states $x_i(k+1)$ are unknown. However, assume the states of the neural network are approximately equal to the desired states at every time instant, then $x_i(k+1)$ can be replaced by $x_i^d(k+1)$. The goal of the neural network is to realize a mapping that approximates the inverse dynamics $(g_i(\cdot))$ of the system. Define the cost function:

$$J = \frac{1}{2}[x_i^d(k) - x_i(k)]^2,\tag{8.3}$$

where $x_i^d(k)$ is the desired system state at time k, $x_i(k)$ is the output of the neural network:

$$\hat{u}_i(k) = NN_i(X, W_{1i}, W_{2i}),\tag{8.4}$$

where $NN_i(\cdot)$ is the functional mapping of the neural network, X is a vector representing all system states, and W_{i1} and W_{i2} are the weight vectors in the output and hidden layers of the neural network. Without the loss of generality, we assume the neural networks used is a network with one hidden layer, and the output layer is linear. Ideally, $\hat{u}_i(k)$ should equal to the desired control in equation (8.2). Using the gradient method, we get

$$\begin{aligned}
\Delta W_{1i} &= -\xi\frac{\partial J}{\partial W_{1i}}\\
&= -\xi\frac{\partial J}{\partial \hat{u}_i(k)}\frac{\partial \hat{u}_i(k)}{\partial W_{1i}}\\
&= -\xi(x_i^d(k) - x_i(k))\sum_{j=1}^{n}\left\{\frac{\partial x_i(k)}{\partial \hat{u}_j(k)}\frac{\partial \hat{u}_j(k)}{\partial W_{1i}}\right\}.
\end{aligned}\tag{8.5}$$

In learning algorithm in equation (8.5), the derivative $\partial\hat{u}_i/\partial W_{1i}$ is easy to get based on the input-output relation of the neural network. However, if the system plant is unknown, then $\partial x_i(k)/\partial \hat{u}_j(k)$ is difficult to know.

In [192], a very simple method for incorporating human knowledge into neural network control has been suggested. It was argued that although the derivative of the system state with respect to the control input is unknown, the relationship between change of the control input and the change of the system state can usually be known. For example, simple knowledge such as "If the system control increases, then the system state also increase," or "If the system control increases, the system state decreases". That is to say, the derivative of the system state with respect to its input can be replaced with the sign function. Thus, the learning algorithm in equation (8.5) can be rewritten as follows:

$$\Delta W_{1i} = -\xi(x_i^d(k) - x_i(k))\sum_{j=1}^{n}\left\{\mathrm{sign}(\frac{\partial x_i(k)}{\partial \hat{u}_j(k)})\frac{\partial \hat{u}_j(k)}{\partial W_{1i}}\right\}.\tag{8.6}$$

However, it has been found that the replacement of the derivative with its sign function is too simple. Besides, in multivariable control, the value of the derivative of the system state $x_i(k)$ with respect to different input $u_j(k)$ varies a lot. For this reason, equation (8.6) can be rewritten as:

$$\Delta W_{1i} = -\xi(x_i^d(k) - x_i(k)) \sum_{j=1}^{n} \left\{ k_{ij} \cdot \text{sign}(\frac{\partial x_i(k)}{\partial \hat{u}_j(k)}) \frac{\partial \hat{u}_j(k)}{\partial W_{1i}} \right\}, \qquad (8.7)$$

where k_{ij} is a constant that can be estimated using the derivative $\partial x_i(k)/\partial \hat{u}_j(k)$ and the absolute value of the current state error

$$e_i(k) = |x_i^d(k) - x_i(k)| \qquad (8.8)$$

with the help of a number of heuristic fuzzy rules such as:

If $\frac{\Delta x_i}{\Delta u_j}$ is *Small*, then k_{ij} is *Small*;

If $\frac{\Delta x_i}{\Delta u_j}$ is *Medium*, then k_{ij} is *Medium*;

If $\frac{\Delta x_i}{\Delta u_j}$ is *Large* and e_i is *Large*, then k_{ij} is *Small*;

If $\frac{\Delta x_i}{\Delta u_j}$ is *Large* and e_i is *Small*, then k_{ij} is *Medium*;

In a study on the control of CSTR reactor [116], it has been shown that this kind of simple knowledge will improve the control performance significantly.

8.3 Fuzzy Knowledge Incorporation By Regularization

8.3.1 Knowledge Representation with Fuzzy Rules

While data collection is expensive in many real-world problems, *a priori* knowledge about the behavior of a system is often available, i.e., from experts, that cannot be described by data. This kind of knowledge can be represented with fuzzy rules using linguistic terms that are used by humans. Thus, heuristics and expert knowledge, which are the most common forms of *a priori* knowledge, can be incorporated into neural networks to improve the learning quality when training data is sparse.

As an example, we deduce a couple of fuzzy rules for a planar robot manipulator with two links. From Figures 8.3, 8.4 and 8.5, and from "common sense", the following nine fuzzy rules can be extracted, which will be used later to complement the conventional purely data-driven learning of the neural network.

1. **If** θ_1 **is small**   and θ_2 is small ,   **then** y is small;
2. **If** θ_1 **is small**   and θ_2 is medium,   **then** y is quite small;
3. **If** θ_1 **is small**   and θ_2 is large,   **then** y is medium;
4. **If** θ_1 **is medium** and θ_2 is small,   **then** y is quite large;
5. **If** θ_1 **is medium** and θ_2 is medium,   **then** y is large;

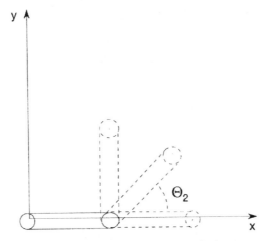

Fig. 8.3. Fuzzy robot kinematics: Θ_1 is small.

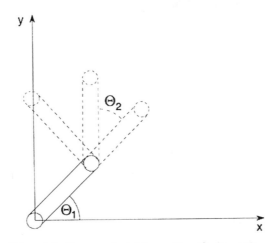

Fig. 8.4. Fuzzy robot kinematics: Θ_1 is medium.

6. If θ_1 is medium and θ_2 is large,   then y is quite large;
7. If θ_1 is large and θ_2 is small,   then y is very large;
8. If θ_1 is large and θ_2 is medium, then y is large;
9. If θ_1 is large and θ_2 is large, then y is medium.

For the formulation of these fuzzy rules three linguistic terms (small, medium and large) are used for the two input variables and six linguistic terms (small, quite small, medium, quite large, large and very large) for the output variable. Note that these fuzzy rules have been obtained solely from observation of the space relationships of the manipulator, without using mathematical analysis of the systems.

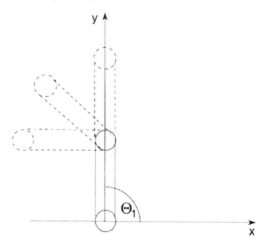

Fig. 8.5. F
uzzy robot kinematics: Θ_1 is **large**.

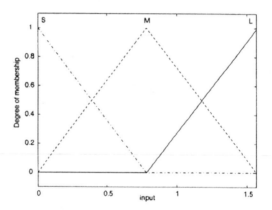

Fig. 8.6. Membership functions for the inputs.

Before these fuzzy rules can be used in neural network learning, proper membership functions must be defined. The definition of the membership functions is based on knowledge from human experts. A straightforward choice is to use the triangular membership functions as shown in Figure 8.6. For the sake of simplicity, we use fuzzy singletons for the output variable; they are approximately set as [0.0, 0.7, 1.0, 1.4, 1.7, 2.0] representing the linguistic terms (**small**, **quite small**, **medium**, **quite large**, **large** and **very large**) under the assumption that the length of each link is 1. Figure 8.7 (a) and (b) show the input-output surface described by the real system and by the fuzzy system.

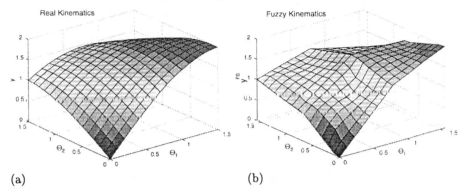

Fig. 8.7. The kinematics of the real system (a) and the fuzzy system (b)

8.3.2 Regularized Learning

In this section, *a priori* knowledge in the form of fuzzy rules will be used
to regularize neural network learning. The idea is to let the neural network
not only learn the data points (provided by the training data), but also the
derivative (provided by the fuzzy system) on these data points. Without the
knowledge about the derivative, the neural network learns only the information at each sample point. If the derivative information at these points
is given, the neural network can additionally learn information about the
behavior of the system between sample points.

In this method, we use the fuzzy derivative for the regularization primarily
to provide extra knowledge to the neural network. Similar to the conventional
regularization algorithms, we add an extra term to the quadratic error function:

$$J = E + \lambda \Omega, \tag{8.9}$$

where, E is the conventional quadratic error function:

$$E = (y^{NN} - y^t)^2, \tag{8.10}$$

where y^{NN} is the output of the neural network and y^t is the target output. In
equation 8.9, λ is the regularization coefficient, which is usually much smaller
than 1 and has to be set manually. Ω is the regularization term describing
the difference between the partial derivatives of the neural network and the
target partial derivatives calculated from the fuzzy rule system:

$$\Omega = \sum_{i=1}^{n} \left(\frac{\partial y^{NN}}{\partial x_i} - \frac{\partial y^{FS}}{\partial x_i} \right)^2, \tag{8.11}$$

where i is the number of inputs.

The MLP neural network model has been discussed in Chapter 2. It is
recalled that an MLP network with one hidden layer can be described as
follows:

$$y^{NN} = \sum_{j=1}^{H} v_j f\left(\sum_{i=1}^{n} w_{ij} x_i\right), \tag{8.12}$$

where H is the number of hidden nodes. Note that the activation function for the output node is linear, whereas the activation function for the hidden neurons is given by the following logistic function:

$$f(z) = \frac{1}{1 + \exp(-z)}. \tag{8.13}$$

Now we can derive the learning algorithm according to the cost function in equation 8.9 using the gradient method. The differentiation of equation 8.9 with respect to the output weights v_j, results in

$$\frac{\partial J}{\partial v_j} = \frac{\partial E}{\partial v_j} + \lambda \frac{\partial \Omega}{\partial v_j} \tag{8.14}$$

In equation (8.14), the first term on the right side is the standard gradient term for neural network learning and will not be re-described here. For the second term, we get:

$$\frac{\partial \Omega}{\partial v_j} = \sum_{i=1}^{n} \left(\frac{\partial y^{NN}}{\partial x_i} - \frac{\partial y^{FS}}{\partial x_i}\right) \frac{\partial(\partial y^{NN}/\partial x_i)}{\partial v_j} \tag{8.15}$$

$$= \sum_{i=1}^{n} \left(\frac{\partial y^{NN}}{\partial x_i} - \frac{\partial y^{FS}}{\partial x_i}\right) \frac{\partial D}{\partial v_j},$$

where:

$$\frac{\partial D}{\partial v_j} = f'(.)w_{ij}. \tag{8.16}$$

Similarly, the learning algorithm for the input weights w_{ij} can be obtained:

$$\frac{\partial \Omega}{\partial w_{ij}} = \sum_{i=1}^{n} \left(\frac{\partial y^{NN}}{\partial x_i} - \frac{\partial y^{FS}}{\partial x_i}\right) \frac{\partial D}{\partial w_{ij}}, \tag{8.17}$$

$$\frac{\partial D}{\partial w_{ij}} = v_j f'(.) + v_j w_{ij} f''(.) x_i, \tag{8.18}$$

where $f'(.)$ and $f''(.)$ are the first and second derivatives of the activation function, thus

$$f'(.) = f(1 - f), \tag{8.19}$$

$$f''(.) = (1 - 2f)f(1 - f). \tag{8.20}$$

In order to calculate the derivative error in equation (8.11), we have to calculate the derivative of the neural network with respect to its inputs x_i:

$$\frac{\partial y^{NN}}{\partial x_i} = \sum_{j=1}^{H} v_j \frac{\partial f}{\partial x_i} \tag{8.21}$$

$$= \sum_{j=1}^{H} v_j f'(.) w_{ij}.$$

There are two different possibilities to calculate the derivative of the fuzzy system. The first approach is to use the analytical form of the fuzzy system and to apply the partial derivative to it. Consider a fuzzy system with N fuzzy rules:

$$y^{FS} = \frac{\sum_{j=1}^{N} r^j y^j}{\sum_{j=1}^{N} r^j}, \tag{8.22}$$

$$r^j = T_{i=1}^{n}(A_{ij}(x_i)), \tag{8.23}$$

where $T(\cdot)$ is the T-norm. In this algorithm, we assume that multiplication is used for the fuzzy intersection, and Gaussian functions are used for fuzzy membership functions:

$$A_{ij}(x_i) = \exp(-b_{ij}(x_i - a_{ij})^2). \tag{8.24}$$

Thus, the derivative of the output of the fuzzy system with respect to the input x_i can be calculated as follows:

$$\frac{\partial y^{FS}}{\partial x_i} = \frac{\partial y^{FS}}{\partial r_j} \frac{\partial r_j}{\partial x_i}, \tag{8.25}$$

where,

$$\frac{\partial y^{FS}}{\partial r_j} = \frac{y_j - \sum_{i=1}^{N} r_j y_j}{(\sum_{i=1}^{N} r_j)^2}, \tag{8.26}$$

$$\frac{\partial r_j}{\partial x_i} = -2(x_i - a_{ij})r_j. \tag{8.27}$$

If the fuzzy membership functions are triangular functions, and if the fuzzy intersection is the minimum, the derivation of the derivative of the fuzzy system becomes a little more complicated.

An alternative approach is to calculate the derivative numerically, which is computationally very simple:

$$\frac{\partial y^{FS}}{\partial x_i} \approx \frac{y^{FS}(x_i + \Delta x_i) - y^{FS}(x_i)}{\Delta x_i}, \tag{8.28}$$

where Δx_i should be selected appropriately.

In the simulation studies, the proposed method will be applied to the problem of neural network learning of the robot kinematics introduced above.

8.4 Fuzzy Knowledge as A Related Task in Learning

8.4.1 Learning Related Tasks

Learning related tasks (also called *multi-task learning* or *learning to learn*) [225] has found to be helpful both in human and machine learning. When a child starts learning, it never learns a single, isolated complex task [29]. On the contrary, a child learns several related tasks at the same time: walk, run, grasp, throw and so on.

In machine learning systems, several topics have been investigated to find out how related tasks can help learning, including *network reuse* and *knowledge transfer, lifelong learning, continual learning* or *multi-task learning*. The most essential idea behind these learning methods is twofold. First, it should be easier to learn several related tasks than to learn a single task. Second, the learner should be able to gain information from learning other related tasks to improve the performance on the new tasks.

Although the idea of learning related tasks simultaneously is compelling, in practice it is very difficult to define a class of related tasks. It is also hard to find out whether two tasks are related. One interesting way is to use multiple representations and define different error metrics [225]. In fact, network ensembles and *constructive learning* (bagging and boosting) may fall in this category. Another useful approach to addressing this problem is to use *a priori* knowledge that approximately describes the original task as a new related task. This idea will be implemented in the next section and has proved to be useful.

In the multi-task learning, a related task that is used to help the neural network learning is sometimes called *catalyst* or *hint*, which has been proposed in [211] to improve the learning performance of feed-forward neural networks. In [1] it has been shown that by introducing hints into neural networks, the effective VC dimension of the problem will be reduced. The VC dimension [19, 230] indicates an upper bound for the number of examples needed by a learning process.

8.4.2 Fuzzy Knowledge as A Related Task

In this section, we use the fuzzy model built on heuristics as a a related task to be learned by the neural network to add knowledge into the network and thus to improve the learning performance.

The basic scheme for multi-task learning using fuzzy rule based *a priori* knowledge is shown in Figure 8.8. Here the output y of the output neuron is required to approximate the target output y^t. In addition, the output y_c of the augmented neuron is required to approximate the output of the fuzzy system, which is similar to the target output and serves as a related task. The output of the fuzzy systems is referred to as y_c^t. During neural network learning, these two output neurons are trained simultaneously. It has been

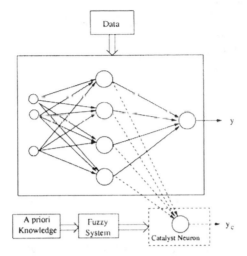

Fig. 8.8. Fuzzy system for multi-task learning.

shown that by adding the extra catalyst task, the network can adapt the weights in the hidden layer more easily for the original task represented by y^t.

Of course, the main problem is to find an appropriate, related task represented by y_c^t. A possible choice for y_c^t is a simplified version of the original output target y^t. Fuzzy systems designed by experts which mainly capture the principles behind a certain task are therefore a sensible definition for y_c^t. We will use the fuzzy system in the previous section as a related task for the neural network. This represents a second way to incorporate *a priori* knowledge into the neural network and to support the standard network learning.

8.5 Simulation Studies

In order to show the effectiveness of the two knowledge incorporation methods, we will implement them for learning robotic kinematics using neural networks. The robot considered in this simulation is the two-link planar manipulator as discussed previously. The length of each link is set to 1 and the angle of each link varies between $[0, \frac{\pi}{2}]$. All samples are generated randomly. The employed neural network is a feed-forward network with one hidden layer and 10 hidden nodes and is trained with the standard error back-propagation algorithm. The activation function for the hidden nodes is the logistic function and the output node is linear.

In each run, two data sets are randomly generated, one training set with 11 samples and the other with 101 samples. For both cases, a test data set consisting of 200 uniformly distributed samples is also generated. It is noticed that the first case represents a small data set that does not contain sufficient information on the target function, whereas the second data set represents a large data set and does have enough information for the network to learn the target function.

The weights of the network with and without knowledge incorporation are initialized randomly in the interval $[-0.05, 0.05]$. All results presented are averaged over twenty different runs.

8.5.1 Regularized Learning

Firstly, we consider the situation in which the available learning samples are insufficient. The eleven samples are used for training the network and 9000 learning cycles are carried out. Figure 8.9(a) shows the training and test errors with and without the fuzzy derivative regularization averaged over 20 runs. We observe that by incorporating knowledge from the fuzzy system, the learning speed is significantly increased and the final error after 9000 learning cycles is considerably reduced. The difference between the final errors (with and without knowledge) on the test set is larger than that on the training set, which agrees well with the assumption that the incorporation of non-data based knowledge can enhance the generalization ability of the network.

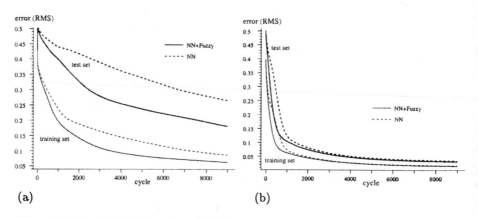

(a)   (b)

Fig. 8.9. Regularized learning: (a) Training and test error for 11 data samples averaged over 20 runs; (b) Training and test error for 101 data samples averaged over 20 runs.

If the available data are sufficient, as in Figure 8.9 (b) (101 samples), the final error is hardly reduced, however, the effect on the learning speed remains. The learning speed increase becomes more evident in Figure 8.10, which shows the first 1500 cycles of Figure 8.9 (b). For example, when no fuzzy knowledge is incorporated, it takes about 800 cycles for the neural network to reach an error of 0.15 on the test data. When fuzzy knowledge is incorporated, it takes about 500 cycles to reach the same error on the test data. In real-time applications, the improvement of the learning speed of the neural network can be essential and even imperative if neural networks are to be employed at all.

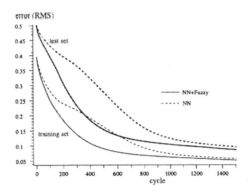

Fig. 8.10. Regularized learning. The training and the test error for 101 samples during the first 1500 learning cycles (the first 1500 cycles of Figure 8.9).

8.5.2 Multi-task Learning

In this section, we will show that fuzzy knowledge can be used as a catalyst function in neural network learning. An output node and the corresponding weights are added to the neural network as shown in Figure 8.8. This node is removed after the network training is finished.

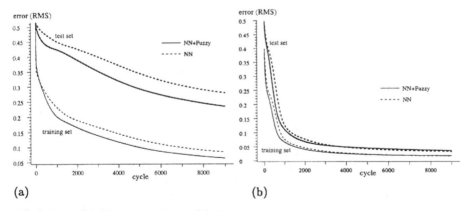

(a)   (b)

Fig. 8.11. Multi-task learning: (a) Training and test error for 11 data samples averaged over 20 runs; (b) Training and test error for 101 data samples averaged over 20 runs.

Similar to the regularized learning, two cases will be considered. In the first case, 11 examples, and in the second cases, 101 examples are generated randomly to train the *task neuron* of the neural network. The same number of examples (identical input) are generated by the fuzzy rule system in order to train the *catalyst neuron*. The training and test results are shown in Figure 8.11 (a) and (b) again averaged over 20 runs. The results are similar to those for using the regularization technique. For the small data set, the network with knowledge incorporation performs better, both with respect to the learning speed and the final error. Again the decrease of the error on the test set indicates improvement of the generalization ability of the network. In the case of the larger data set, Figure 8.11 (b), the findings are similar to those in the last section. Figure 8.12 (a) highlights the increase of the learning speed of the network in multi-task learning.

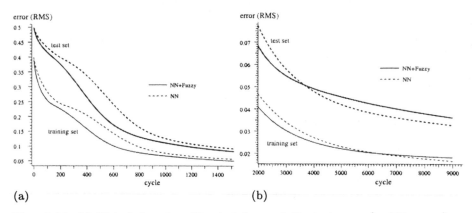

(a)   (b)

Fig. 8.12. Multi-task learning. The training and the test error for 101 samples during the (a) first 1500 learning cycles (the first 1500 cycles of Figure 8.11) and (b) during learning cycles 2000–9000.

A closer observation of the curves at later cycles, as shown in Figure 8.12 (b), reveals that the error curves for both the training and the test data sets for the network with and without knowledge incorporation intersect after 6500 cycles and 3500 cycles, respectively. Since the effect also occurs on the test data, it is unlikely that over-training of the feed-forward neural network is responsible. An alternative explanation is that, due to its *fuzzy* nature, the additional knowledge itself is not error free. Therefore, after a certain number of cycles, the network performance depends on the accuracy of the additional knowledge. If this is indeed the case, then a possible extension would be to reduce the influence of the additional knowledge with increasing number of cycles. In this case, the additional knowledge would be exploited at the start of the training process as a means of speeding up the adaptation and it would be removed in later periods when the fine tuning of the network should depend solely on the data, which are supposed to more accurate than the *a priori* knowledge. Of course, for insufficient data sets (Figure 8.11 (a)) this effect cannot occur.

8.6 Summary

Two methods for the incorporation of *a priori* knowledge into learning systems have been introduced in this chapter. Before it is used, the knowledge is represented by fuzzy rules.

In the regularized learning, an additional term is added to the conventional cost function that tries to reduce the derivative error between the fuzzy system and the neural network. In this way, the knowledge in the fuzzy system is transferred to the neural network, which is not included in the training data.

In the second approach, the *a priori* knowledge represented by the fuzzy system is used as a task that is related to the original one. By learning these "two" related tasks, the neural network is able to improve its learning performance significantly.

With the help of a simple simulation study, it is shown that both methods work well with respect to reducing learning error when training data is insufficient and speeding up learning when there are sufficient data.

9. Fuzzy Preferences Incorporation into Multi-objective Optimization

9.1 Multi-objective Optimization and Preferences Handling

9.1.1 Multi-objective Optimization

To facilitate the discussions on multi-objective optimization (MOO), we first give a short review on the definitions related to multiobjective optimization.

Consider a multi-objective optimization problem with k objectives ($f_i, i = 1, 2, ..., k$) and n decision variables ($x_i, i = 1, 2, ..., n$):

$$\mathbf{f}(\mathbf{x}) = (f_1(\mathbf{x}), ..., f_k(\mathbf{x})). \tag{9.1}$$

The target of the optimization is to minimize $f_i(\mathbf{x}), i = 1, 2, ..., k$ subject to

$$g_i(\mathbf{x}) \leq 0, i = 1, 2, ..., m. \tag{9.2}$$

Since the k objectives may be conflicting with each other, it is usually difficult to obtain the global minimum for each objective. Therefore, the target of MOO is to achieve a set of solutions that are Pareto optimal. The related concepts of Pareto dominance, Pareto optimality, Pareto optimal set and Pareto front are defined as follows [231]:

Pareto dominance: A vector $\mathbf{u} = (u_1, ..., u_k)$ is said to dominate a vector $\mathbf{v} = (v_1, ..., v_k)$ (denoted by $\mathbf{u} \preceq \mathbf{v}$) if and only if $u_i \leq v_i, i = 1, 2, ..., k$ and there exists at least one element that $u_i < v_i$.

Pareto optimality: A solution $\mathbf{x}$ is said to be Pareto optimal if and only if there does not exist another solution $\mathbf{x}'$ so that $\mathbf{f}(\mathbf{x})$ is dominated by $\mathbf{f}(\mathbf{x}')$. All the solutions that are Pareto optimal for a given multi-objective optimization problem are called the Pareto optimal set ($\mathcal{P}^\star$).

Pareto front: For a given multi-objective optimization problem and its Pareto optimal set $\mathcal{P}^\star$, the Pareto front ($\mathcal{PF}^\star$) is defined as:

$$\mathcal{PF}^\star = \mathbf{f}(\mathbf{x}) = (f_1(\mathbf{x}), ..., f_k(\mathbf{x})) | \mathbf{x} \in \mathcal{P}^\star. \tag{9.3}$$

There are generally convex and concave Pareto fronts. A Pareto front
($\mathcal{PF}^\star$) is said to be convex if and only if

$$\forall \mathbf{u}, \mathbf{v} \in \mathcal{PF}^\star, \forall \lambda \in (0,1), \exists \mathbf{w} \in \mathcal{PF}^\star : \lambda||\mathbf{u}|| + (1-\lambda)||\mathbf{v}|| \geq ||\mathbf{w}||. \qquad (9.4)$$

On the contrary, a Pareto front is said to be concave if and only if

$$\forall \mathbf{u}, \mathbf{v} \in \mathcal{PF}^\star, \forall \lambda \in (0,1), \exists \mathbf{w} \in \mathcal{PF}^\star : \lambda||\mathbf{u}|| + (1-\lambda)||\mathbf{v}|| \leq ||\mathbf{w}||. \qquad (9.5)$$

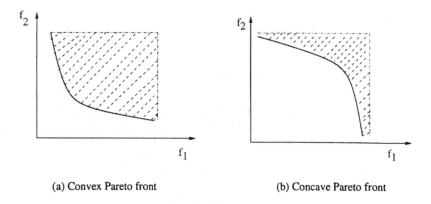

(a) Convex Pareto front   (b) Concave Pareto front

Fig. 9.1. Convex and concave Pareto fronts.

For example, Fig. 9.1(a) is a convex Pareto front and Fig. 9.1(b) is a
concave Pareto front. Of course, a Pareto front can be partly convex and
partly concave, see Fig. 9.2.

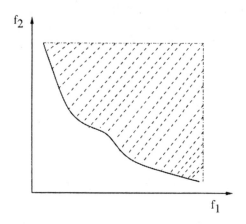

Fig. 9.2. A Pareto front that is partially convex and partially concave.

9.1.2 Incorporation of Fuzzy Preferences

Usually, it is assumed that each objective is of equal importance. This is not always the case in real applications. Very often, a decision-maker has preferences over different objectives. Therefore, finding the complete Pareto-optimal solutions is not the final goal: a decision has to be made on the available alternatives. Usually, a decision is made based on the decision-maker's preferences. This can be done either before, during or after the optimization process takes place.

- **Before optimization.** When there is *a prior* knowledge or preference, one is able to assign a weight for each objective and then combine the weighted objectives into one single scalar function. Of course, it is also possible to incorporate the preferences into Pareto ranking in optimization [42]. In [128], fuzzy preferences are converted into weight intervals that are integrated with the dynamic weighed aggregation to obtain a number of solutions instead of only one.
- **During optimization.** This can usually be realized by an interactive process in which the user is able to interact with the optimizer during optimization. The earliest effort on using interactive approaches to multi-objective optimization might have been be made in [71]. Interactive evolutionary multi-objective optimization is reported in [222] among others. Besides, interactive optimization is also very important for the decomposition of large-scale problems.
- **After optimization.** In most cases, the decision-maker is presented with a set of Pareto-optimal solutions, which might be not globally optimal. Then the decision-maker needs to choose one or a number of solutions based on his preferences. Thus, the goal of the optimization is to obtain a set of Pareto solutions that are complete and globally optimal (the true Pareto front).

As discussed in [42], incorporation of fuzzy preferences can be realized in two ways.

- The weighted sum method. Use the preferences as *a priori* knowledge to determine the weight for each objective and then apply the weights directly to sum up the objectives to a scalar. In this case, only one solution will be obtained.
- The weighted Pareto method. The non-fuzzy weight is used to define a weighted Pareto non-dominance:

$$\mathbf{u} \succeq_w^\tau \mathbf{v} \text{ if and only if } \sum_{i=1}^k w_i I_\geq (\mathbf{u_i}, \mathbf{v_i}) \geq \tau, \tag{9.6}$$

where τ is a real number satisfying $0 < \tau \leq 1$, and

$$I_\geq (\mathbf{u_i}, \mathbf{v_i}) = \begin{cases} 1, \ u_i \geq v_i, \\ 0, \ u_i < v_i, \end{cases} \tag{9.7}$$

and $\sum_{i=1}^{k} w_i = 1$. Note that the weighted dominance reduces to the standard one if $\tau = 1$ and $w_1 = ... = w_k = 1/k$.

In [42], a method for converting linguistic fuzzy preference relations into crisp weights for optimization has been introduced. Since there are three forms of fuzzy models, as discussed Chapter 1, it is necessary to first transform the different models into one uniform model with linguistic preferences. If there is more than one expert, a collective preference has to be obtained.

One weakness when converting fuzzy preferences into real-valued weights is that a lot of information will be lost in the process. In this chapter, we develop a method that converts fuzzy preference relations into crisp weight intervals. Then, the weight intervals are combined with the dynamic weighted aggregation method. In this way, we are able to obtain a number of Pareto solutions instead of only one. Fig. 9.3 provides an overview of the different methods involved in applying fuzzy preferences in multi-objective optimization.

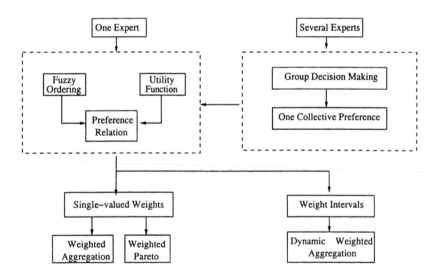

Fig. 9.3. An overview of fuzzy preferences in MOO.

9.2 Evolutionary Dynamic Weighted Aggregation

Evolutionary multi-objective optimization has been widely investigated in the recent years [37, 61]. Generally speaking, there are three main approaches to evolutionary multi-objective optimization, namely, the weighted aggregation approaches, the population-based non-Pareto approaches and the Pareto-based approaches [61]. Currently, the Pareto-based approaches are the most popular ones in the evolutionary computation community [231].

The conventional weighted aggregation based approaches have three main weaknesses. Firstly, aggregation based approaches can provide only one Pareto optimal solution in one run of optimization. Secondly, *a priori* knowledge concerning the system to be optimized is needed to determine the weights for each objective. Thirdly, it has been shown that the conventional weighted aggregation method is unable to deal with multi-objective optimization problems with a concave Pareto front [56].

One effort using the weighted aggregation method for multi-objective optimization (MOO) was reported in [82]. In that work, the weights of the different objectives are encoded in the chromosome to obtain more than one Pareto solution. Phenotypic fitness sharing is used to keep the diversity of the weight combinations and mating restrictions are required so that the algorithm can work properly.

In the next section, a new aggregation based MOO method that is simple but effective will be introduced. This method is called the evolutionary dynamic weighted aggregation (EDWA), see also in [117, 120]. The basic idea in the EDWA is as follows. If the weights for the different objectives are changing during optimization, all the areas of the Pareto front will be reached by the evolutionary algorithm, and if the found solutions are archived, the whole Pareto front can be approximated.

More importantly, it is also shown that the EDWA is able to address MOO problems with a concave Pareto front. Further study on this issue results in a new theory that explains why the EDWA is able to approximate a concave Pareto front whereas the conventional weighted aggregation method fails.

9.2.1 Conventional Weighted Aggregation for MOO

The conventional weighted aggregation (CWA) is a straightforward approach to multi-objective optimization. In this method, the different objectives are summed up to a single scalar with predefined weights

$$F = \sum_{i=1}^{k} w_i f_i(\mathbf{x}), \tag{9.8}$$

where w_i is the non-negative weight for objective $f_i(\mathbf{x}), i = 1, ..., k$. Usually, *a priori* knowledge is needed to specify the weights. During the optimization, the weights are fixed in the conventional weighted aggregation method.

Using this method, only one Pareto optimal solution can be obtained from one run of the optimization algorithm. In other words, if one intends to obtain different Pareto solutions, one has to run the optimization algorithm several times. This is of course not feasible in many real-world problems because it usually takes too much time to run the optimization more than once.

However, efficiency is not the only problem. It was pointed out that the CWA method is not able to obtain the Pareto solutions that are located in the concave region of the Pareto front [56].

9.2.2 Dynamically Weighted Aggregation

To overcome the weaknesses of the conventional weighted aggregation method, three new methods will be presented by extending the conventional weighted aggregation method. The key idea behind the new methods is that the weights should be changed dynamically during optimization.

Random Distribution of Weights. For the sake of clarity, we again consider two-objective problems, although the proposed approach can be extended to problems more than two objectives. In this approach, the different weight combinations are distributed randomly among the individuals. Suppose the population size is P, then the weight combinations can be distributed uniformly among the P individuals in the offspring population. Let

$$w_1^i(t) = \text{random}(P)/P, \tag{9.9}$$
$$w_2^i(t) = 1.0 - w_1^i(t), \tag{9.10}$$

where $i = 1, 2, ..., P$ and t is the index for generation number. Function $random(P)$ generates a uniformly distributed random number between 0 and P. In this way, we can get a uniformly distributed random weight combination (w_1^i, w_2^i) among the individuals, where $0 \leq w_1^i, w_2^i \leq 1$ and $w_1^i + w_2^i = 1$, refer to Fig. 9.4. Notice that the weights are re-generated in every generation.

The extension of the weight change methods in equations (9.9) to optimization problems with more than two objectives is theoretically straightforward. For example, if there are three objectives, equation (9.9) can be rewritten by:

$$w_1^i(t) = \text{random}(P)/P,$$
$$w_2^i(t) = (1.0 - w_1^i(t)) \, \text{random}(P)/P,$$
$$w_3^i(t) = 1.0 - w_1^i(t) - w_2^i(t), \tag{9.11}$$
$$i = 1, 2, \cdots, P.$$

Generation-Based Periodical Change of the Weights. The idea of a uniformly distributed weight combination can straightforwardly be extended to a generation based approach. However, instead of using a randomly distributed weight combination, we use a weight combination that is changed gradually and periodically when the evaluation proceeds. For example, the weights can be changed as follows:

$$w_1(t) = |\sin(2\pi t/F)|, \tag{9.12}$$
$$w_2(t) = 1.0 - w_1(t), \tag{9.13}$$

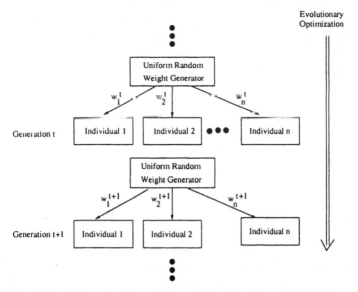

Fig. 9.4. Determination of the weights: Random distribution.

where t is the generation number. This method is termed as the dynamic weighted aggregation (DWA).

We can see from equation (9.12) that $w_1(t)$ changes from 0 to 1 periodically. The frequency of change can be adjusted by F. We found that the results of the algorithm are not very sensitive to F, although it seems reasonable to let the weights change from 0 to 1 twice. In this case, all individuals have the same weight combination in the same generation.

If there are three objectives, the DWA method in equation (9.12) can be modified in the following way.

> Let $t_1 = 0$, $t_2 = 0$, and $t = t_1 + t_2$
> **for** $t_1 = 0$ to $F/2$
> $w_1(t) = |\sin(2\pi t_1/F)|$;
>   **for** $t_2 = 0$ to $F/2$
> $w_2(t) = (1.0 - w_1(t))|\sin(2\pi t_2/F)|$;
> $w_3(t) = 1.0 - w_1(t) - w_2(t)$;
>   **end**;
> **end**;

Note that all weights should be changed smoothly so that the population does not need to jump from one point to another on the Pareto front, which should be avoided if the Pareto front is convex. Furthermore, the number of generations required increases exponentially with the number of objectives if the DWA is used. This is caused by the phenomenon known as "the curse of dimensionality".

Bang-bang Weighted Aggregation. The bang-bang weighted aggregation (BWA) can be seen as a special case of dynamic weighted aggregation. A bang-bang change of weights can be realized in the following way for a two-objective minimization problem:

$$w_1(t) = \text{sign}(\sin(2\pi t/F)) \tag{9.14}$$
$$w_2(t) = 1.0 - w_1, \tag{9.15}$$

where t is the generation index and F is the frequency of the weight change. It is clear that F should be large enough to allow the evolutionary algorithm to move from one stable point to another.

9.2.3 Archiving of Pareto Solutions

For all the three algorithms described above, the population is not able to maintain all found Pareto solutions, although the evolutionary algorithm moves along the Pareto front dynamically. Therefore, it is necessary to record the Pareto solutions that have been found so far. The idea of building a Pareto archive is inspired from [139], although we use a different method for the maintenance of the archive. The pseudo-code for building the archive is as follows:

For each individual o in the offspring population

If o dominates an individual in the parent population p, and

if o is not dominated by any solutions in the archive, and

if o is not similar to any solutions in the archive, and

if the archive is not full, add o to the archive

else if o dominates any solution a in archive, replace a with o

else if any solution in archive a_1 dominates another solution a_2, replace a_2 with o

else discard o

else discard o

For each solution in the archive

If solution a_1 dominates a_2

remove a_2

The similarity is measured by the Euclidean distance in the fitness space.

9.2.4 Simulation Studies

Test Functions. To evaluate the effectiveness of the proposed algorithms, simulations are carried out on six test functions. The first three test functions are taken from [257, 139] and the fourth test function is a combination of test functions F_2 and F_3 so that its Pareto front is partially convex and partially concave. Test functions F_5 has a discontinuous Pareto front.

- F_1 is the Schaffer's function [195]. Its Pareto front is convex and uniform.

$$f_1 = \frac{1}{n} \sum_{i=1}^{n} x_i^2, \tag{9.16}$$

$$f_2 = \frac{1}{n} \sum_{i=1}^{n} (x_i - 2.0)^2. \tag{9.17}$$

- The second test function (F_2) is the first function in [257], which has a convex Pareto front:

$$f_1 = x_1, \tag{9.18}$$

$$g(x_2, ..., x_n) = 1.0 + \frac{9}{n-1} \sum_{i=2}^{n} x_i, \tag{9.19}$$

$$f_2 = g \times (1.0 - \sqrt{f_1/g}). \tag{9.20}$$

where $x_i \in [0, 1]$.
- The third test function (F_3) is the second function in [257], which has a concave Pareto front:

$$f_1 = x_1, \tag{9.21}$$

$$g(x_2, ..., x_n) = 1.0 + \frac{9}{n-1} \sum_{i=2}^{n} x_i, \tag{9.22}$$

$$f_2 = g \times (1.0 - (f_1/g)^2). \tag{9.23}$$

where $x_i \in [0, 1]$.
- The fourth test function (F_4) is derived from F_2 and F_3, which has a Pareto front that is neither purely convex nor purely concave:

$$f_1 = x_1, \tag{9.24}$$

$$g(x_2, ..., x_n) = 1.0 + \frac{9}{n-1} \sum_{i=2}^{n} x_i, \tag{9.25}$$

$$f_2 = g \times (1.0 - \sqrt[4]{f_1/g} - (f_1/g)^4), \tag{9.26}$$

where $x_i \in [0, 1]$.
- The fifth test function (F_5) is the third function in [257], whose Pareto front consists of a number of separated convex parts:

$$f_1 = x_1, \tag{9.27}$$

$$g(x_2, ..., x_n) = 1.0 + \frac{9}{n-1} \sum_{i=2}^{n} x_i, \tag{9.28}$$

$$f_2 = g \times (1.0 - \sqrt{f_1/g} - (f_1/g)\sin(10\pi f_1)), \tag{9.29}$$

where $x_i \in [0, 1]$.

- F_6 has been studied in [62], which has a concave Pareto front.

$$f_1 = 1. - \exp\left\{-\sum_{i=1}^{n}\left(x_i - \frac{1}{\sqrt{n}}\right)^2\right\}, \qquad (9.30)$$

$$f_2 = 1. - \exp\left\{-\sum_{i=1}^{n}\left(x_i + \frac{1}{\sqrt{n}}\right)^2\right\}. \qquad (9.31)$$

In the simulations, the standard evolution strategy (μ, λ)-ES [199] has been employed. μ is set to 15 and λ is set to 100 and 150 generations are run in each optimization.

We first try to use the conventional weighted aggregation to obtain the Pareto front. As we pointed out before, we have to run the evolutionary algorithm more than once if we intend to get more than one solution. For test functions F_1, F_2 and F_3, 20 times are run, and for test function F_4, 40 times are run. Since the Pareto solutions are not uniformly distributed on the Pareto front corresponding to a uniformly distributed weight combination, finer weight change is needed to obtain the solutions in the convex region of the Pareto front in function F_4. In all the simulations, the dimension is set to $n = 2$.

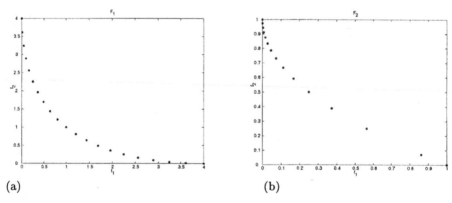

(a)   (b)

Fig. 9.5. CWA for F_1 and F_2. The results are collected from 20 runs of the optimization.

The results on F_1 and F_2 is given in Fig. 9.5. Since both functions are convex, the CWA based approach is able to obtain different Pareto solutions with different weights. However, distribution of the solutions from F_2 is not uniform, although the distribution of the weights is uniform.

Fig. 9.6 provides the results on F_3 and F_4. Since the Pareto front of F_3 is concave, we can only obtain two solutions at $(f_1, f_2) = (0, 1)$ and $(f_1, f_2) = (1, 0)$. For F_4, the solutions in the convex region are obtained and those in the concave region are not obtained.

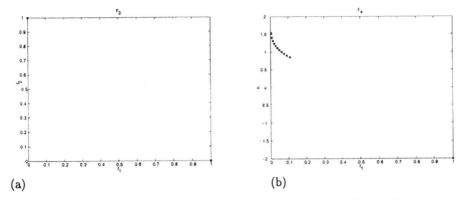

(a)   (b)

Fig. 9.6. CWA for F_3 and F_4. 20 runs are carried out for F_3 and 40 runs are carried for F_4.

Bang-bang Weighted Aggregation. The simulations are carried out on F_1, F_3, F_4 and F_5 using the BWA method, which represent convex, concave, mixed and discrete Pareto fronts, respectively.

The results on F_1 is shown in Fig. 9.7 (a). It can be seen that the BWA method may have poorer performance than the DWA method for convex functions. The result on F_3 is given in Fig. 9.7 (b). It turned out that the BWA method also works for MOO problems with a concave Pareto front and from the figure, we see that a quite complete Pareto front has been obtained.

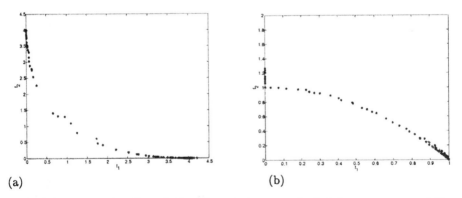

(a)   (b)

Fig. 9.7. BWA for F_1 and F_3. The results are collected from one runs of the optimization. A Pareto solution is denoted by '*' and a solution in the current population by '+'.

The results from F_4 and F_5 are presented in Fig. 9.8 (a) and (b). Note that the most part of the Pareto front for F_4 is concave with a small convex section. It can be seen that BWA has also successfully obtained the Pareto front.

The results of using BWA on the discontinuous Pareto front is very interesting. Although the different sections are not connected with each other, the evolutionary algorithm is able to traverse through all the sections of the Pareto front. The dynamic process is shown by a snapshot during optimization Fig. 9.9.

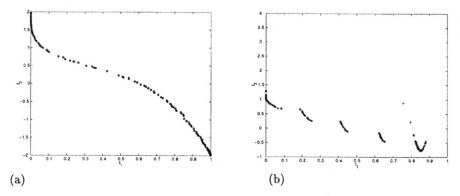

(a)   (b)

Fig. 9.8. BWA for F_4 and F_5. The results are collected from one runs of the optimization. A Pareto solution is denoted by '*' and a solution in the current population by '+'.

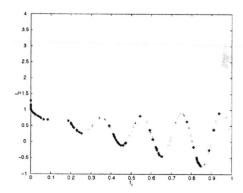

Fig. 9.9. A snapshot showing how the evolutionary algorithm is moving from one section of the Pareto front to another. A '*' denotes a Pareto-optimal solution in the archive, a green '+' represents a solution in the current parent population and a blue '+' means a solution in the offspring population.

Dynamic Weighted Aggregation. In the following, we intend to show that dynamic weighted aggregation can not only obtain the Pareto solutions in one run, but also is able to obtain the solutions in the concave region of the Pareto front.

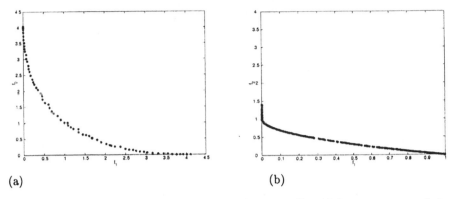

(a)   (b)

Fig. 9.10. DWA for F_1 and F_2. The results are collected from one runs of the optimization. A Pareto solution is denoted by '$*$' and a solution in the current population by '$+$'.

In Fig. 9.10, the results on F_1 and F_2 using DWA are provided. It is clearly shown that in one run of the optimization, the Pareto solutions have been obtained. Similarly, a quite complete Pareto set is also obtained for test functions F_3 and F_4, whose Pareto front is not convex. The result on F_5 with a discontinuous Pareto front is shown in Fig. 9.12.

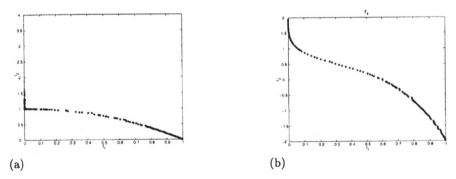

(a)   (b)

Fig. 9.11. DWA for F_3 and F_4. The results are collected from one runs of the optimization. A Pareto solution is denoted by '$*$' and a solution in the current population by '$+$'.

The results on test function F_6 is presented in Fig. 9.13, which is also a concave test function.

It can be seen from the figures that the proposed methods have successfully obtained the Pareto front for all test problems, no matter whether their Pareto front is convex, concave or non-convex.

From the simulation results, the following two important observations can be made concerning the DWA method.

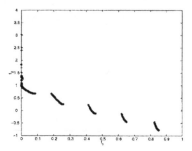

Fig. 9.12. DWA for F_5 with a discontinuous Pareto front. A Pareto solution is denoted by '*' and a solution in the current population by '+'.

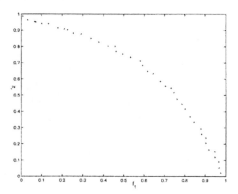

Fig. 9.13. Achieved Pareto front for test function F_6.

- If the Pareto front is convex, the population will first converge to a point on the Pareto front. To which point the population converges depends on the problem difficulty (convergence speed) and the speed of weight change. Once the population has reached the Pareto front, it will keep moving on it as the weights change. The speed of movement can be controlled by the speed of weights change. Fig.9.14 shows the trace of the parent population during optimization of the test function F_1 in the first 100 generations represented by the mean of the population.
- If the Pareto front is concave, the population will converge to one of the two ends of the concave Pareto front. After that, the population will remain on this Pareto optimal point until the weights are changed to a threshold value, which is determined by the characteristics of the Pareto front. Then, the population will move along or close to the Pareto front very quickly to the other end of the Pareto front. To illustrate this, the first 100 moves of the parent population during optimization of test function F_6 are presented in Fig. 9.15. From the figure, it can be seen that the population converges to the Pareto point $(f_1, f_2)=(0.98,0)$ in a few generation. As the weights

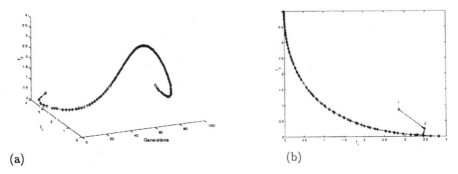

(a)   (b)

Fig. 9.14. Trace of the population for a convex problem (F_1) using DWA. (a) Moves of the population with time (generation); (b) Moves of the population on the fitness space. The arrows denote the direction of the movement and the numbers the generation index.

change, the population remains on the same point until generation 29. Then, the population moves along the Pareto front quickly to the other end of the Pareto front, where $(f_1, f_2) = (0.0, 0.98)$. Again, the population remains on this point until generation 91.

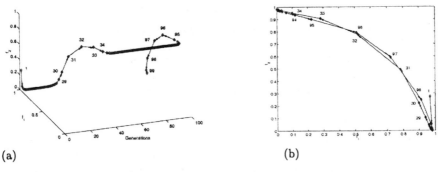

(a)   (b)

Fig. 9.15. Trace of the population for a concave problem (F_6). (a) Moves of the population with time (generation); (b) Moves of the population in the fitness space. The arrows denote the direction of the movement and the numbers the generation index.

9.2.5 Theoretical Analysis

Through a variety of test functions, it has been shown that the EDWA method works very well for approximation of both convex and concave Pareto fronts. However, an analytical proof of the method may be difficult, if not impossible. Nevertheless, it is still of great significance to gain a deeper insight into the working mechanisms of evolutionary weighted aggregation method. In the following, we first show schematically the reasons why the conventional weighted method based multiobjective optimization has several weaknesses and why the EDWA method is able to overcome the shortcomings. Then, we show the global convexity empirically for all the test functions. After that, we present the results from some test functions that the Pareto-optimal solution is a natural local attractor in the fitness space. Finally, some remarks are made concerning the implementation of the EDWA method.

Stable Pareto-optimal Solutions. To explain the reasons why the EDWA method is able to overcome the weaknesses of the conventional weighted aggregation method, we first discuss schematically when a Pareto-optimal solution is stable using the conventional weighted aggregation method. A Pareto optimal solution is defined as stable if this solution is capturable using the conventional weighted aggregation method for a given weight combination. We use the word "stable" instead of "capturable" because we want to show the capturabililty of a solution using weighted aggregation methods by making an analogue to the stability of a ball (point) on a surface under the gravity. In this analogue, a solution on the Pareto-optimal front is treated as a ball (point) on the surface, the Pareto front is the surface, the selection pressure is the gravity, and the change of the weights is equivalent to the rotation of the surface. For the sake of clarity, we discuss bi-objective problems. We first consider a convex problem. In Fig. 9.16(a), point A is the stable solution on the convex Pareto front corresponding to a weight combination, say $(w_1,w_2) = (w_1^A,w_2^A)$. For this weight combination, point B is a not stable. Therefore, it cannot be obtained with the weighted aggregation method using this weight combination. However, if the Pareto front rotates θ_B degrees, see Fig. 9.16(b), which now corresponds to a weight combination of $(w_1,w_2) = (w_1^B,w_2^B)$, point B becomes stable and point A non-stable. To gradually decrease the weight for f_1 and increase the weight for f_2 is the same as to rotate the Pareto front counter-clockwise. Therefore, if the Pareto front is not rotated, which corresponds to a fixed weight combination in the conventional weighted aggregation method, only one solution can be obtained in one run of optimization. On the contrary, if the Pareto front is rotated gradually during optimization, which corresponds to the DWA method, the ball will go through every point on the Pareto front. This means that all solutions on a convex Pareto front can be obtained in one run of optimization, although the found solutions need to be stored in an archive [117].

If the Pareto front is concave, the situation is much different. It is seen that given the weight combination $(w_1,w_2) = (w_1^A,w_2^A)$, point A is stable and point B is unstable, refer to Fig. 9.17(a). When the Pareto front rotates for θ_B degrees, see Fig. 9.17(b), neither point A nor point B is stable. Instead, point C becomes stable. Therefore, point C instead of point B is obtained if the weights are changed from $(w_1,w_2) = (w_1^A,w_2^A)$ to $(w_1,w_2) = (w_1^B,w_2^B)$. It is straightforward to see that if the Pareto front is concave, only one of its two ends (point A and C in Fig. 9.17) can be a stable solution, no matter how the weights are changed. This is the reason why the conventional weighted aggregation method can obtain only one of the two stable solutions for any given weight combinations. Nevertheless, if the Pareto front is rotated, the ball is still able to move along the Pareto front from one stable point to the other, which means that the DWA method is able to get a number of the Pareto-optimal solutions even if the Pareto front is concave.

The phenomena observed in the simulations can be easily explained according to the above discussion. That is, once the population reaches any point on a convex Pareto front, it will keep moving on the Pareto front when the weights change gradually. In contrast, when a concave Pareto front rotates, the population will remain on the stable solution until this solution becomes unstable. Then the population will move along or close to the Pareto front from the unstable solution to the stable one.

It should be emphasized that our analysis does not conflict with the conclusions drawn in [43]. The reason is that although the points in the concave region of a Pareto front are not stable solutions and thus cannot be obtained by the conventional weighted aggregation method, they are reachable.

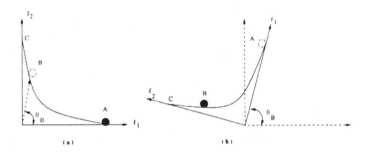

Fig. 9.16. The case of a convex Pareto front. (a) The stable point is A; (b) After the Pareto front rotates θ_B degrees, the stable point becomes B.

One remaining question is: when does a stable solution on a concave Pareto front become unstable and when does the population begin to move? Obviously, there is a dividing point (denoted as point D in Fig. 9.17) on any concave Pareto fronts. The stable solution in Fig. 9.17(a) becomes unstable only when the rotation angle is larger than θ_D. This agrees well with the movement of the population shown in Fig. 9.15(a).

Global Convexity. To make the mechanism explained practical, one very important condition must be satisfied. That is, if an individual intends to move from one Pareto point (A) to a neighboring point (B) in a few steps, the solutions in the parameter space $(a$ and $b)$ should also be in the neighborhood, as illustrated in Fig. 9.18. We term this the "neighborhoodness" condition. This kind of neighborhoodness has also been observed in combinatorial multiobjective problems [20]. Meanwhile, since the Pareto-optimal

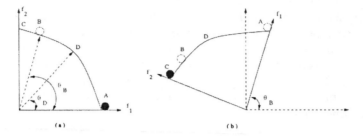

Fig. 9.17. The case of a concave Pareto front. (a) The stable point is A; (b) After the Pareto front rotates θ_B degrees, the stable point is C, instead of point B. Point D is the dividing point, which means that point A becomes unstable until the Pareto front rotates θ_D degrees.

solutions are rather concentrated in the fitness space, i.e., the Pareto front, the solutions in the parameter space should also locate in a very small region of the parameter space. These are actually the two main aspects of the global convexity in multiobjective optimization.

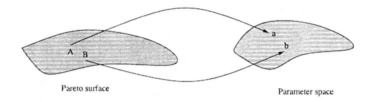

Pareto surface   Parameter space

Fig. 9.18. Mapping from the fitness space to the parameter space.

It is very interesting and surprising to verify that for all the test functions, the solutions in the parameter space are really concentrated in very small regions that can be described by a piecewise linear function. We call it the *definition function* of the Pareto front, which is a function in the parameter space that defines the Pareto front in the fitness space.

The solutions of function F_1 obtained in the simulations are shown in Fig. 9.19 (a). These solutions can be described by a linear definition function. Using the least square method, the following linear model can be obtained, which is also illustrated in Fig. 9.19 (a):

$$x_2 = 0.0296 + 0.981x_1. \tag{9.32}$$

It means that the Pareto front can be directly obtained using the definition function in equation (9.32). This is verified in Fig. 9.19 (b).

Similarly, the following definition function can be obtained for F_2:

$$x_2 = 0. \tag{9.33}$$

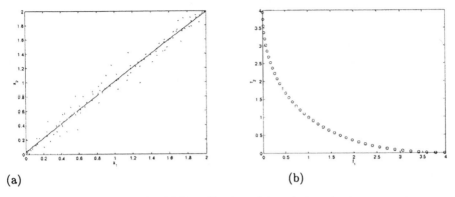

(a)   (b)

Fig. 9.19. Test function F_1. (a) Distribution of the solutions in the parameter space and the estimated definition; (b) Pareto front reconstructed from the estimated definition function.

The distributions of the solutions, the estimated definition function and the Pareto front reconstructed from the definition function are shown in Fig. 9.20.

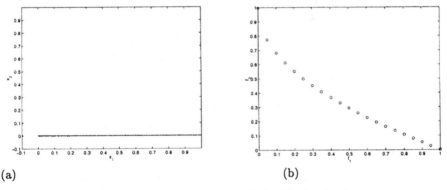

(a)   (b)

Fig. 9.20. Test function F_2. (a) Distribution of the solutions in the parameter space and the estimated definition; (b) Pareto front reconstructed from the estimated definition function.

Now we try to construct an approximate model for the definition function of the Pareto front for test functions F_3 and F_4. Using the solutions obtained in optimization, the following linear model has been obtained for F_3:

$$x_2 = -0.0397 + 0.9035x_1, -0.70 \le x_1 \le 0.70. \tag{9.34}$$

The distribution of the solutions in the parameter space and the estimated definition function are shown in Fig. 9.21(a), whereas the Pareto front reconstructed from the estimated definition function is illustrated in Fig. 9.21(b). It can be seen that the Pareto front reconstructed from the estimated definition function is comparable to the one obtained in optimization (Fig. 9.13).

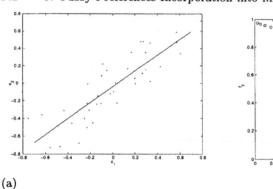
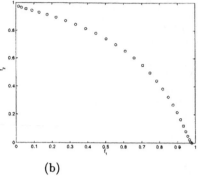

(a)   (b)

Fig. 9.21. Test function F_3. (a) Distribution of the solutions in the parameter space and its estimated model; (b) Pareto front reconstructed from the estimated definition function.

Similarly, approximate models for the definition function of F_4 can also be constructed. From the solutions, the following piece-wise linear model has been obtained, which consists of four sections:

$$\text{section 1: } x_2 = -0.002 + 0.999x_1; \ 0.45 \le x_1 \le 2.0, \tag{9.35}$$
$$\text{section 2: } \quad x_2 = 1.684 + 0.80x_1; \quad 1.35.0 \le x_1 \le -0.55, \tag{9.36}$$
$$\text{section 3: } \quad x_2 = -1.47 + 0.60x_1; \quad 0.45.0 \le x_1 \le 1.35, \tag{9.37}$$
$$\text{section 4: } x_2 = -0.002 + 0.999x_1; \ -2.0 \le x_1 \le -0.55. \tag{9.38}$$

The estimated definition model and the reconstructed Pareto front are given in Fig. 9.22 (a) and (b), respectively. It is noticed that section BC on the Pareto optimal front can be reconstructed either from section 2 or from section 3 of the definition function.

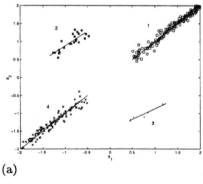
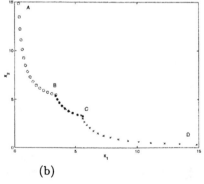

(a)   (b)

Fig. 9.22. Test function F_4. (a) Distribution of the solutions in the parameter space and its estimated model; (b) Pareto front reconstructed from the estimated definition function.

Until now, we have shown that the Pareto-optimal solutions of all the test functions are gathered in a very small fraction of the parameter space. Moreover, they can be described by a piecewise linear function. In the following, we will show empirically that all the test functions exhibit the second aspect of global convexity, i.e., if two solutions are in the neighborhood in the fitness space, then they are also in the neighborhood in the parameter space.

The neighborhoodness can be measured by calculating the Manhattan distance between the two points. Given two vectors $u = \{u_1, .., u_n\}$ and $v = \{v_1, .., v_n\}$, the Manhattan distance between them is defined by:

$$d(u, v) = \sum_{l=1}^{n} |u_l - v_l|. \qquad (9.39)$$

With the help of the definition function of the Pareto front, it is straightforward to calculate the Manhattan distance between two Pareto-optimal solutions in the parameter space, as well as in the fitness space. The results from the four test functions are presented in Figures 9.23, 9.24, 9.25 and 9.26, respectively. Since we are investigating the neighborhoodness of multiobjective optimization, the maximal distance between two Pareto points shown in the figures is 0.5 for F_1 and F_4, and 0.1 for F_2 and F_3, which are scaled according to the range of the Pareto fronts.

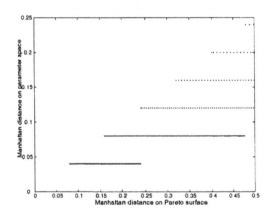

Fig. 9.23. Manhattan distance for test function F_1.

From the figures, it is confirmed that for any two neighboring points on the Pareto front, their corresponding solutions in the parameter space are also in the neighborhood. Generally speaking, the smaller the distance on the Pareto front, the smaller the distance in parameter space, although this relationship is not monotonous, i.e., given the same distance on the Pareto front, the distance on the parameter space may vary a little.

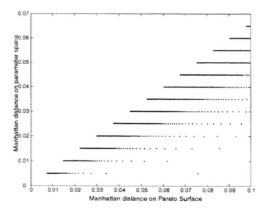

Fig. 9.24. Manhattan distance for test function F_2.

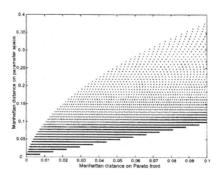

Fig. 9.25. Manhattan distance for test function F_3.

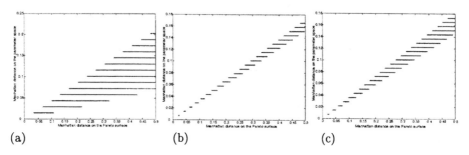

(a)   (b)   (c)

Fig. 9.26. Manhattan distance for test function F_4. (a) Section 1 and 4; (b) Section 2; (c) Section 3.

The global convexity of multiobjective optimization is essential for the success of the EDWA. Only with these conditions being satisfied, the population is able to move along (or close to) the Pareto front when the weights change during the optimization, once the population reaches any point on the Pareto front.

Local Attractor. The property of the global convexity is an important prerequisite for the success of the proposed method. Besides, another phenomenon, namely, the Pareto front is sometimes a natural attractor in the fitness space, is helpful in achieving the Pareto front. Of course, this phenomenon is not a necessary condition for working mechanism of our method.

The phenomenon that the Pareto front is a local attractor in the fitness space can be shown by investigating the mapping from the parameter space to the fitness space in multiobjective problems. In evolution strategies, mutation is the main search mechanism, which is usually implemented by adding a random number from a normal random distribution to the objective parameters. It is therefore very important to see how a normal random distribution is changed by the mapping from the parameter space onto the fitness space. Without the loss of generality, we consider the two-dimensional two-objective optimization problems. The distribution in the fitness space can be derived from the distribution in the parameter space as follows, provided that the fitness function is a one-to-one function:

$$\int_{f_1}^{f_1+\Delta f_1} \int_{f_2}^{f_2+\Delta f_2} g(f_1', f_2')df_2'df_1' = \int_{x_1}^{x_1+\Delta x_1} \int_{x_2}^{x_2+\Delta x_2} f(x_1', x_2')dx_2'dx_1',$$

$$(9.40)$$

where $g(f_1, f_2)$ is the probability density distribution (pdf) in the fitness space, and $f(x_1, x_2)$ is a normal random distribution in the parameter space:

$$f(x_1, x_2) = \frac{1}{2\pi\sigma_1\sigma_2}e^{-\frac{(x_1-\mu_1)^2}{2\sigma_1^2}} e^{-\frac{(x_2-\mu_2)^2}{2\sigma_2^2}}, \qquad (9.41)$$

where, $f(x_1, x_2)$, σ_i and μ_i $(i = 1, 2)$ are the probability density function, the standard deviation and the mean, respectively. Equation 9.40 can be further written by:

$$g(f_1, f_2) = \frac{1}{|J|}f(x_1, x_2), \qquad (9.42)$$

where J is the Jacobian matrix. Refer to [178] for more discussions on this property of evolutionary multiobjective optimization.

Fig. 9.27 shows the probability distributions in the fitness space for F_1 that are mapped from normal distributions in the parameter space, where the darker the color, the higher the probability. It can be seen that given a normal distribution in the parameter space, its corresponding distribution in the fitness space is deviated toward the Pareto front. Therefore, the Pareto front is a local attractor in the fitness space in a probability sense. We call it a local attractor because this phenomenon can be observed only if the center of the distribution in the parameter space is relatively close to or on the curve of the definition function.

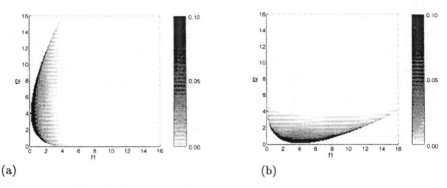

(a)   (b)

Fig. 9.27. The distributions $g(f_1, f_2)$ in the fitness space for test function F_1 generated from a normal distribution in the parameter space on (a) $(x_1, x_2)=(0,0)$, (b) $(x_1, x_2)=(2,2)$. The standard deviation of distribution $f(x_1, x_2)$ is $(\sigma_1, \sigma_2) = (1, 1)$.

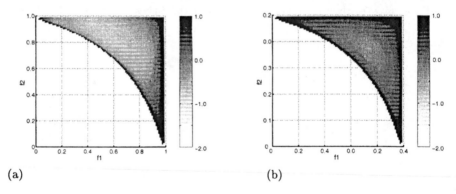

(a)   (b)

Fig. 9.28. The distributions $g(f_1, f_2)$ in the fitness space for test function F_6 generated from a normal distribution in the parameter space on (a) $(x_1, x_2)=(-\frac{1}{\sqrt{2}}, -\frac{1}{\sqrt{2}})$, (b) $(x_1, x_2)=(0,0)$. The standard deviation of distribution $f(x_1, x_2)$ is $(\sigma_1, \sigma_2) = (1, 1)$.

Similar phenomena have been observed for test functions F_6, which are shown in Fig. 9.28.

In contrast, this phenomenon is not observed for test function F_2. Fig. 9.29 shows that the distributions in the fitness space almost remain to be normal ones. Nevertheless, the proposed algorithm has obtained the Pareto front successfully, which again implies that the local attractor property is very helpful but not a prerequisite for the proposed method.

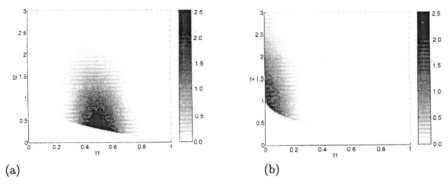

Fig. 9.29. The distributions $g(f_1, f_2)$ in the fitness space for test function F_2 generated from a normal distribution in the parameter space on (a) $(x_1, x_2) = (0.5, 0)$, (b) $(x_1, x_2) = (0, 0)$. The standard deviation of distribution $f(x_1, x_2)$ is $(\sigma_1, \sigma_2) = (0.01, 0.01)$.

9.3 Fuzzy Preferences Incorporation in MOO

9.3.1 Converting Fuzzy Preferences into Crisp Weights

Given $m = 1, ..., L$ experts and their preference relation P^m, where P^m is a $k \times k$ matrix with p_{ij} denoting the preference of objective o_i over objective o_j, $i, j = 1, ..., k$. Then, based on the group decision making method introduced in Chapter 1, the individual preferences can be combined into a single collective preference P^c. Each element of the preference matrix is defined by one of the following linguistic terms:

> {*Much more important (MMI), more important (MI), equally important (EI), less important (LI), much less important (MLI)*}.

For the sake of simplicity, we omit the superscript c indicating the collective preference in the following text. In converting the linguistic terms into concrete weights, it is necessary to use the following evaluations [42] to replace the linguistic preferences p_{ij} in the preference matrix:

$$\text{If } a \text{ is } \textit{Much less important} \text{ than } b \Longrightarrow p_{ij} = \alpha, p_{ji} = \beta,$$
$$\text{If } a \text{ is } \textit{Less important} \text{ than } b \Longrightarrow p_{ij} = \gamma, p_{ji} = \delta,$$
$$\text{If } a \text{ is } \textit{Equally important} \text{ as } b \Longrightarrow p_{ij} = \epsilon, p_{ji} = \epsilon.$$

The value of the parameters needs to be assigned by the decision-maker and the following conditions should be satisfied in order not to lose the interpretability of the linguistic terms:

$$\alpha < \gamma < \epsilon = 0.5 < \delta < \beta, \tag{9.43}$$
$$\alpha + \beta = 1 = \gamma + \delta. \tag{9.44}$$

Consider an MOO problem with six objectives $\{o_1, o_2, ..., o_6\}$ [42] . Suppose among these six objectives, o_1 and o_2, o_3 and o_4 are *equally important*. Thus we have four classes of objectives: $c_1 = \{o_1, o_2\}, c_2 = \{o_3, o_4\}, c_3 = \{o_5\}$ and $c_4 = \{o_6\}$. Besides, we have the following preference relations:

c_1 is much more important than c_2;
c_1 is more important than c_3;
c_4 is more important than c_1;
c_3 is much more important than c_2.

From these preferences, it is easy to get the following preference matrix:

$$P = \begin{bmatrix} EI & MMI & MI & LI \\ MLI & EI & MLI & MLI \\ LI & MMI & EI & LI \\ MI & MMI & MI & EI \end{bmatrix}. \tag{9.45}$$

From the above fuzzy preference matrix, we can get the following concrete relation matrix R:

$$R = \begin{bmatrix} \epsilon & \beta & \delta & \gamma \\ \alpha & \epsilon & \alpha & \alpha \\ \gamma & \beta & \epsilon & \gamma \\ \delta & \beta & \delta & \epsilon \end{bmatrix}. \tag{9.46}$$

Based on this relation matrix, the weight for each objective can be obtained by:

$$w(o_i) = \frac{S(o_i, R)}{\sum_{i=1}^{k} S(o_i, R)}, \tag{9.47}$$

where

$$S(o_i, R) = \sum_{j=1, j \neq i}^{k} p_{ij}. \tag{9.48}$$

For the above example, we have

$$w_1 = w_2 = \frac{2 - \alpha}{8 + 2\alpha}, \tag{9.49}$$

$$w_3 = w_4 = \frac{3\alpha}{8 + 2\alpha}, \tag{9.50}$$

$$w_5 = \frac{1 - \alpha + 2\gamma}{8 + 2\alpha}, \tag{9.51}$$

$$w_6 = \frac{3 - \alpha - 2\gamma}{8 + 2\alpha}. \tag{9.52}$$

Using this method, the fuzzy preferences on the objectives can be converted into a single-valued weights, which can then be applied to the conventional weighted aggregation method to achieve one Pareto optimal solution.

9.3.2 Converting Fuzzy Preferences into Weight Intervals

It is noticed that the value of the weights is not unique because the value of the parameters, α, β, γ and δ is not unique. Therefore, it makes more sense to convert the fuzzy preferences into weight intervals.

To show how the value of the parameters effects that of the weights, an experiment is carried out on the example. Fig. 9.30 and Fig. 9.31 show the change of the weights with the change of the parameters.

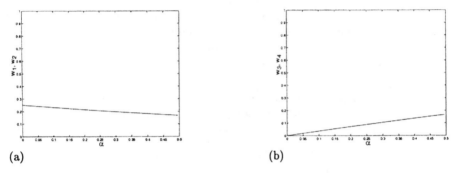

(a)   (b)

Fig. 9.30. Change of the weights with the change of the parameters. (a) w_1, w_2; (b) w_3, w_4.

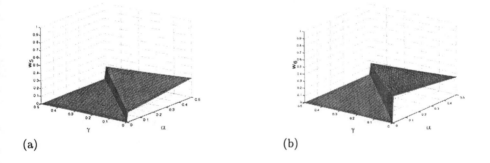

(a)   (b)

Fig. 9.31. Change of the weights with the change of the parameters. (a) w_5; (b) w_6.

It is seen that the fluctuation of the weights is not very significant when the parameters change. However, it is still impractical to apply one single weight combination in optimization from a number of fuzzy preferences. One approach to address this problem is to embody the change of the weights in the optimization. This can be realized with the dynamic weighted aggregation proposed in this Chapter. Suppose the maximal and minimal value of a weight is w^{max} and w^{min} when the parameters change, then the random weight in equation (9.9) can be modified as follows:

$$w_1^i(t) = w_1^{i,min} + (w_1^{i,max} - w_1^{i,min}) * random(P)/P, \qquad (9.53)$$
$$w_2^i(t) = 1.0 - w_1^i(t), \qquad (9.54)$$

Similarly, the DWA can be modified as follows to find out the preferred Pareto solutions:

$$w_1(t) = w_1^{min} + (w_1^{max} - w_1^{min}) * |\sin(2\pi t/F)|. \qquad (9.55)$$

In this way, the evolutionary algorithm can achieve a set of Pareto solutions reflected by the fuzzy preferences. However, since the DWA cannot control the movement of the individuals if the Pareto front is concave, fuzzy preferences incorporation into MOO using the DWA is applicable to convex Pareto fronts only, whereas the RWA works for both convex and concave fronts.

To illustrate how this method works, examples on two-objective optimization problems are presented in the following text. In the simulations, we consider two different fuzzy preferences:

    1. *Objective 1 is more important than objective 2.*
    2. *Objective 1 is less important than objective 2.*

For the first preference, we can get the following preference matrix:

$$P = \begin{bmatrix} 0.5 & \delta \\ \gamma & 0.5 \end{bmatrix}, \qquad (9.56)$$

where $0.5 < \delta < 1$ and $0 < \gamma < 0.5$. Therefore, the weights for the two objectives using the RWA are:

$$w_1^i(t) = 0.5 + 0.5 * random(P)/P, \qquad (9.57)$$
$$w_2^i(t) = 1.0 - w_1^i(t). \qquad (9.58)$$

Similarly we can get the weights for the second preference:

$$w_1^i(t) = 0 + 0.5 * random(P)/P, \qquad (9.59)$$
$$w_2^i(t) = 1.0 - w_1^i(t). \qquad (9.60)$$

The weights of the objectives using the DWA can be obtained in a similar way. For the first preference, we have:

$$w_1(t) = 0.5 + 0.5 * |\sin(2\pi t/F)|, \qquad (9.61)$$
$$w_2(t) = 1.0 - w_1(t). \qquad (9.62)$$

For the second preference, the weights change as follows:

$$w_1(t) = 0 + 0.5 * |\sin(2\pi t/F)|, \qquad (9.63)$$
$$w_2(t) = 1.0 - w_1(t). \qquad (9.64)$$

Simulation results are carried out on the first three test functions in [120], where F_1 and F_2 have a convex Pareto front, and F_3 has a concave Pareto front. The simulation results on F_1 and F_2 are given in Figures 9.32 and 9.33 using the RWA method, and in Figures 9.34 and 9.35 using the DWA method.

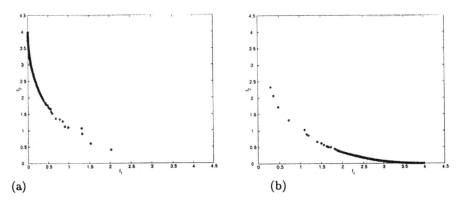

(a)   (b)

Fig. 9.32. RWA results on F_1. (a) f_1 is more important than f_2; (b) f_1 is less important than f_2

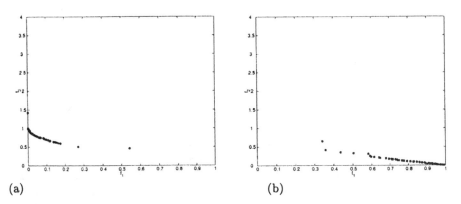

(a)   (b)

Fig. 9.33. RWA results on F_2. (a) f_1 is more important than f_2; (b) f_1 is less important than f_2

It can be seen that the performance of the RWA and the DWA are similar on the two test functions with a convex Pareto front. However, the performance on the test function F_3, which has a concave Pareto front, is very different. The performance of the DWA is quite bad, see Fig. 9.36(b), whereas the performance of the RWA is acceptable, refer to Fig. 9.36(a).

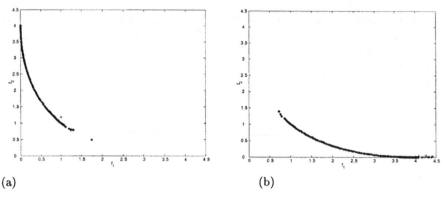

(a)   (b)

Fig. 9.34. DWA results on F_1. (a) f_1 is more important than f_2; (b) f_1 is less important than f_2

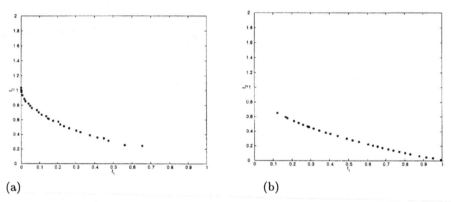

(a)   (b)

Fig. 9.35. DWA results on F_2. (a) f_1 is more important than f_2; (b) f_1 is less important than f_2

9.4 Summary

Fuzzy preference incorporation into evolutionary multi-objective optimization is discussed in this chapter. Instead of converting the fuzzy preferences into one concrete weight combination, a method is suggested to convert the fuzzy preferences into weight intervals. This weight intervals are applied to the evolutionary dynamic weighted aggregation method (EDWA) to get the preferred Pareto solutions.

The EDWA is a new method to deal with MOO problems based on weighted aggregation. It has shown that the EDWA works well MOO problems, no matter the Pareto front is convex or concave. This leads to a new theory for the weighted aggregation method for MOO, which has been empirically verified on a number of test functions.

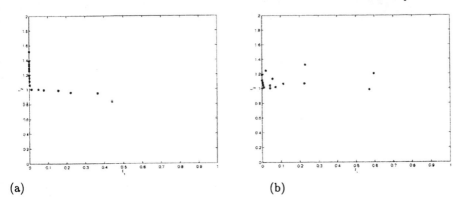

(a)   (b)

Fig. 9.36. Results on F_3 for the preference "f_1 is more important than f_2". (a) RWA; (b) DWA.

References

1. Y.S. Abu-Mostafa. Hints and the VC dimension. *Neural Computation*, 5(2):278–288, 1993.
2. R. Andrews, T. Diederich, and A. Tickle. A survey and critique of techniques for extracting rules from trained artificial neural networks. *Knowledge Based Systems*, 8:373–389, 1995.
3. K.J. Arrow. *Social Choice and Individual Values*. John Wiley, New York, 1951.
4. T. Bäck. Parallel optimization of evolutionary algorithms. In *Parallel Problem Solving from Nature*, pages 418–427. Springer, 1994.
5. T. Bäck. *Evolutionary Algorithms in Theory and Practice*. Oxford University Press, Oxford, 1996.
6. T. Bäck and H. Schwefel. An overview of evolutionary algorithms for parameter optimization. *Evolutionary Computation*, 1(1):1–23, 1993.
7. P. V. Balakrishnan, M.C. Cooper, V.S. Jacob, and P.A. Lewis. A study of the classification capabilities of neural networks using unsupervised learning: A comparison with k-means clustering. *Psychometrika*, 59:509–525, 1994.
8. J.-F.M. Barthelemy. Approximation concepts for optimimum structural design - a review. *Structural Optimization*, 5:129–144, 1993.
9. A.G. Barto. Neuron-like adaptive elements that can solve difficult learning control problems. *IEEE Transactions on Systems, Man, and Cybernetics,*, 13:834–846, 1983.
10. R. Belew. When both individuals and populations search: Adding simple learning to the genetic algorithm. In J.D. Schaffer, editor, *Proceedings of 3rd International Conference on Genetic Algorithms*. Morgan Kaufmann, 1989.
11. R. Bellman. *Adaptive Control Process: A Guided Tour*. Princeton University Press, 1961.
12. J.M. Benitez, J.L. Castro, and I. Requena. Are artificial neural networks black boxes. *IEEE Transactions on Neural Networks*, 8(5):1156–1164, 1997.
13. P.J. Bentley. "Evolutionary, my dear watson"- investigating committee-based evolution of fuzzy rules for detection of suspicious insurance claims. In *Proceedings of the Genetic and Evolutionary Computation Conference*, pages 702–709, Las Vegas, Nevada, 2000. Morgan Kaufmann.
14. P.J. Bentley. Representations are more important than algorithms: Why evolution needs embryology. In *Proceedings of the Fourth International Conference on Adaptive Computing in Design and Manufacture*, Plymouth, UK, 2000.
15. H.R. Berenji and P. Khedkar. Learning and tuning of fuzzy logic controllers through reinforcements. *IEEE Trasactions on Neural Networks*, 3:724–740, 1992.
16. A.P. Bertsekas and J.N. Tsitsiklis. *Neuro-dynamic Programming*. Athena Scientific, Belmont, MA., 1996.
17. J. A. Biles. Genjam: A genetic algorithm for generating jazz solos. In *Proceedings of International Computer Music Conference*, pages 131–137, 1994.

18. C. M. Bishop. *Neural Networks for Pattern Recognition.* Oxford University Press, Oxford, UK, 1995.
19. A. Blumer, A. Ehrenfeucht, D. Haussler, and M. Warmuth. Learnability and the VC dimension. *Journal of ACM*, 36:929–965, 1989.
20. P.C. Borges and M.P. Hansen. A basis for future successes in multiobjective combinatorial optimization. Technical Report IMM-REP-1998-8, Department of mathematical Modeling, Technical University of Denmark, 1998.
21. G. Box. Evolutionary operation: A method for increasing industrial productivity. *Applied Statistics*, 6:81–101, 1957.
22. J. Branke. Creating robust solutions by means of evolutionary algorithms. In *Proceedings of Parallel Problem Solving from Nature*, Lecture Notes in Computer Science, pages 119–128. Springer, 1998.
23. J. Branke, C. Schmidt, and H. Schmeck. Efficient fitness estimation in noisy environment. In L. Spector et al, editor, *Proceedings of Genetic and Evolutionary Computation*, pages 243–250, San Francisco, CA, July 2001. Morgan Kaufmann.
24. H. Braun and P. Zagorsky. ENZO-M- A hybrid approach for optimizing neural networks by evolution and learning. In Y. Davidor, H.-P Schwefel, and R. Männer, editors, *Proceedings of the 3rd International Conference on Parallel Problem Solving from Nature*, pages 440–451, Jerusalem, Israel, 1994.
25. L. Breiman. Bagging predictors. *Machine Learning*, 24(2):123–240, 1996.
26. A. Buczak, Y. Jin, T. Houshang, and M. Jafari. Genetic algorithm based sensor network optimization for target tracking. In C.H. Dagli, A. Buczak, J. Ghosh, M. Embrechs, and O. Ersoy, editors, *Intelligent Engineering Systems Through Artificial Neural Networks*, volume 9, pages 349–354, 1999.
27. L. Bull. On model-based evolutionary computation. *Soft Computing*, 3:76–82, 1999.
28. E. Cantu-Paz. A summary of research on parallel genetic algorithms. Technical report, Illinois Genetic Algorithms Laboratory, University of Illinois at Urnana-Champaign, Urbana-Champaign, 1995.
29. R. Caruana. Learning many related tasks at the same time with back-propagation. In G. Tesauro, D. Touretzky, and T. Leen, editors, *Advances in Neural Information Processing Systems*, pages 657–664. Morgan Kaufmann, 1995.
30. R.A. Caruana and J.D. Schaffer. Representation and hidden bias: Gray versus binary coding in genetic algorithms. In J. Laird, editor, *Proceedings of 5th Conference on Machine Learning*, pages 153–161, Ann Arbor, MI, 1988. Morgan Kaufmann.
31. J. L. Castro, C. J. Mantas, and J. Benitez. Interpretation of artificial neural networks by means of fuzzy rules. *IEEE Transactions on Neural Networks*, 13(1):101–116, 2002.
32. A. Cechin, U. Epperlein, B. Koppenhoefer, and W. Rosenstiel. The extraction of sugeno fuzzy rules from neural networks. In M. Verleysen, editor, *Proceedings of the European Symposium on Artificial Neural Networks*, pages 49–54, Bruges, Belgium, 1996.
33. D. L. Chester. Why two hidden layers are better than one. In L. Erlbaum, editor, *International Joint Conference on Neural Networks*, pages 265–268, Washington D.C., 1990.
34. F. Chiclana, F. Herrera, and E. Herrera-Viedma. Integrating three representation models in fuzzy multi-purpose decision making based on fuzzy preference relations. Technical Report DECSAI-96106, Department of Computer Science and Artificial Intelligence, University of Granada, May 1996.

35. M.-Y. Chow, S. Altrug, and H.J. Trussell. Heuristic constraints enforcement for training of and knowledge extraction from a fuzzy/neural architecture - Part I: Foundations. *IEEE Transactions on Fuzzy Systems*, 7(2):143–150, 1999.

36. P. Coates. Using genetic programming and L-systems to explore 3D design worlds. In R. Junge, editor, *CAAD Future'97*, Munich, 1997. Kluwer Academic Press.

37. C.A.C. Coello. Constraint handling through a multi-objective optimization technique. In A. Wu, editor, *Proceedings of the 1999 Genetic and Evolutionary Computation Conference. Workshop Program*, pages 117–118, Orlando, FL, 1999.

38. C.A.C. Coello. Handling preferences in evolutionary multi-objective optimization: A survey. In *Proceedings of IEEE Congress on Evolutionary Computation*, pages 30–37, Piscataway, New Jersey, 2000. IEEE.

39. J. Cohen and I Stewart. *The Collapse of Chaos*. Penguin Books, 1994.

40. O. Cordon and F. Herrera. A hybrid genetic algorithm-evolution strategy process for learning fuzzy logic controller knowledge base. In F. Herrera and J.L. Verdegay, editors, *Genetic Algorithms and Soft Computing*, pages 251–278. Physica-Verlag, 1996.

41. T. Cover and J. Thomas. *Elements of Information Theory*. John Wiley and Sons, Inc, 1991.

42. D. Cvetkovic and I. Parmee. Use of preferences for GA-based multi-objective optimization. In *Proceedings of 1999 Genetic and Evolutionary Computation Conference*, pages 1504–1510, San Francisco, California, 1999. Morgan Kaufmann.

43. I. Das and J.E. Dennis. A closer look at drawbacks of minimizing weighted sum of objectives for Pareto set generation in multicriteria optimization problems. *Structural Optimization*, 14(1):63–69, 1997.

44. D. Dasgupta. Incorporate redundancy and gene activation mechanisms in genetic search for adapting to non-stationary environments. In L. Chambers, editor, *Practical Handbook of Genetic Algorithms*. CRC Press, 1995.

45. K. De Jong. *Analysis of the behavior of a class of genetic adaptive systems*. PhD thesis, University of Michigan, Ann Arbor, 1975.

46. T. Dean, J. Allen, and Y. Aloimono. *Artificial Intelligence: Theory and Practice*. Addison-Wesley, 1995.

47. G. Deco, W. Finnoff, and H.G. Zimmermann. Unsupervised mutual information criterion for elimination of over-training in supervised multi-layer networks. *Neural Computation*, 7:86–107, 1995.

48. J. Dennis and V. Torczon. Managing approximate models in optimization. In N. Alexandrov and M. Hussani, editors, *Multidisciplinary design optimization: State-of-the-art*, pages 330–347. SIAM, 1997.

49. Y.-S. Ding, H. Ying, and S.-H. Shao. Necessary conditions on minimal system configuration for general miso fuzzy systems as universal approximators. *IEEE Transactions on Systems, Man, and Systems*, 30:857–864, 2000.

50. D. Dubois and H. Prade. *Fuzzy Sets and Systems: Theory and Applications*. Academic Press, New York, 1980.

51. D. Dubois and H. Prade. What are fuzzy rules and how to use them. *Fuzzy Sets and Systems*, 84:169–185, 1996.

52. R. Durbin and D.E. Rumelhart. Product units: A computationally powerful and biologically plausible extension to backpropagation networks. *Neural Computation*, 1:133–142, 1989.

53. C. Elkan. The paradoxical success of fuzzy logic. In *Proceedings of the Eleventh National Conference on Artificial Intelligence*, pages 698–703. AIAA Press, 1993.

54. S.E. Fahlman and C. Lebiere. The cascade-correlation learning architecture. In *Advances in Neural Information Processing Systems*, volume 2, pages 524–532. Morgan Kaufmann, 1990.

55. J.M. Fitzpatrick and J.J. Grefenstette. Genetic algorithms in noisy environments. *Machine Learning*, 3:101–120, 1988.

56. P.J. Fleming. Computer aided control systems using a multi-objective optimization approach. In *Proc. IEE Control'85 Conference*, pages 174–179, Cambridge, U.K., 1985.

57. J.C. Fodor and M. Roubens. *Fuzzy Preference Modeling and Multi-criteria Decision Support*. Kluwer, Dordrecht, 1994.

58. T.C. Fogarty. Varying the probability of mutation in the genetic algorithm. In J.D. Schaffer, editor, *Proceedings of International Conference on Genetic Algorithms*, pages 104–109, Fairfax, VA, 1989. Morgan Kanfmann.

59. D. Fogel. *Evolving Artificial Intelligence*. PhD thesis, University of California, San Diego, 1992.

60. L. Fogel, J. Owens, and M. Walsh. *Artificial Intelligence through Simulated Evolution*. John Wiley, New York, 1966.

61. C. M. Fonseca and P. J. Fleming. Multi-objective optimization. In Th. Bäck, D. B. Fogel, and Z. Michalewicz, editors, *Evolutionary Computation*, volume 2, pages 25–37. Institute of Physics Publishing, Bristol, 2000.

62. C.M. Fonseca and P.J. Fleming. Genetic algorithm for multiobjective optimization: Formulation, discussion and generalization. In S. Forrest, editor, *Proceedings of the Fifth International Conference on Genetic Algorithms*, pages 416–423, San Mateo, 1993. Morgan Kaufmann.

63. C.M. Fonseca and P.J. Flemming. On the performance assessment and comparison of stochastic multi-objective optimizers. In H.-M. Voigt, W. Ebeling, I. Rechenberg, and H.-P Schwefel, editors, *Parallel Problem Solving from Nature*, volume IV, pages 584–593, Berlin, 1996. Springer.

64. P. Frasconi, M. Gori, M. Maggini, and G. Soda. Unified intergration of explicit knowledge and learning by example in recurrent networks. *IEEE Transactions on Knowledge and data Engineering*, 7(2):340–346, 1995.

65. A.S. Fraser. Simulation of genetic systems by automatic digital computers. I. Introduction. *Australian Journal of Biological Science*, 10:484–491, 1957.

66. E. Freud. The structure of decoupled nonlinear system. *International Journal of Control*, 21:443–450, 1975.

67. Y. Freund and R. Schapire. Experiments with a new boosting algorithm. In L. Saitta, editor, *Proceedings of the Thirteenth International Conference on Machine Learning*, pages 148–156, San Francisco, CA, 1996. Morgan Kaufmann.

68. G. Friedmann. Digital simulation of an evolutionary process. *General Systems Yearbook*, 4:171–184, 1959.

69. A. Geist, A. Beguellin, J. Dongarra, W. Jiang, R. Manchek, and V. Sunderam. *Parallel Virtual Machine: A Guide to Networked and Parallel Computing*. The MIT Press, 1994.

70. S. Geman, E. Bienenstock, and R. Doursat. Neural networks and the bias/variance dilemma. *Neural Computation*, 4:1–58, 1992.

71. A. Geoffrion, J. Dyer, and A. Feinberg. An interactive approach for multi-criterion optimization. *Management Science*, 19:357–368, 1972.

72. D. Goldberg. *Genetic Algorithms in Search, Optimization and machine Learning*. Addison-Wesley, Reading, MA, 1989.

73. D. Goldberg, K. Deb, and B. Korb. Messy genetic algorithms: Motivation, analysis, and first results. *Complex Systems*, 3:493–530, 1989.

74. D.E. Goldberg and K. Deb. A comparative analysis of selection schemes used in genetic algorithms. In *Foundations of Genetic Algorithms*, pages 69–93, San Mateo, 1990. Morgan Kaufmann.

75. J. Graf and W. Banzhaf. Interactive evolution of images. In J.R. McDonnell, R.G. Reynolds, and D. Fogel, editors, *Proceedings of 4th Conference on Evolutionary Programming*, pages 53–65, Cambridge, MA., 1995. MIT Press.

76. P.J. Green and B.W. Silverman. *Non-parametric Regression and Generalized Linear Models: A Roughness Penalty Approach*. Chapman & Hall, London, 1994.

77. J. Grefenstette. Parallel adaptive algorithms for function optimization. Technical Report CS-81-19, Vanderbilt University, Nashville,TN, 1981.

78. D.E. Grierson and W.H. Pak. Optimal sizing, geometrical and topological design using a genetic algorithm. *Structural Optimization*, 6(3):151–159, 1993.

79. P. Grosso. *Computer simulations of genetic adaptation: Parallel subcomponent interaction in a multilocus model*. PhD thesis, University of Michigan, 1985.

80. F. Gruau. Genetic synthesis of boolean neural networks with a cell rewriting developmental process. In D. Whitley and J. Schaffer, editors, *Proceedings of International Workshop on Combination of Genetic Algorithms and Neural Networks*, pages 55–73. IEEE Press, 1992.

81. T. Hackworth. genetic algorithms: Some effects of redundancy in chromosome. Available at: http://www.cs.ucl.ac.uk/staff/P.Bentley/birkbeck/tim.txt, 1999.

82. P. Hajela and C.-Y. Lin. Genetic search strategies in multi-criterion optimal design. *Structure Optimization*, 4:99–107, 1992.

83. N. Hansen and A. Ostermeier. Completely derandomized self-adaptation in evolution strategies. *Evolutionary Computation*, 9(2):159–195, 2001.

84. N. Hansen, A. Ostermeier, and A. Gawelczyk. On the adaptation of arbitrary normal mutation distributions in evolution strategies: The generating set adaptation. In L.J. Eshelman, editor, *Proceedings of International Conference on Genetic Algorithms*, pages 57–64. Morgan Kaufmann, 1995.

85. I. Harvey. Is there another new factor in evolution? *Evolutionary Computation*, 4(3):311–327, 1997.

86. B. Hassibi and D.G. Stork. Optimal brain surgeon. In S.J. Hanson, J.D. Cowan, and C.L. Giles, editors, *Advances in Neural Information Processing*, volume 5, pages 164–171. Morgan Kaufmann, San Mateo, 1993.

87. F. Herrera, E. Harrera-Viedma, and J.L. Verdegay. On dominance degrees in group decision making with preferences. In *Proceedings of 4th International Workshop on Current Issues in Fuzzy Technologies: Decision Models and Systems*, pages 113–117, Trento, 1994.

88. F. Herrera, E. Herrera-Viedma, and J.L. Verdegay. Direct approach processes in group decision making using linguistic OWA operators. *Fuzzy Sets and Systems*, 79:175–190, 1996.

89. F. Herrera and J.L. Verdegay. On group decision making under linguistic preferences and fuzzy linguistic quantifiers. In B. Bouchon, R. Yager, and L. Zadeh, editors, *Fuzzy Logic and Soft Computing*, pages 173–180. World Scientific, 1995.

90. G.E. Hinton and S.J. Nowlan. How learning can guide evolution. *Complex Systems*, 1:495–502, 1987.

91. F. Hoffmann and G. Pfister. Learning of a fuzzy control rule base using messy genetic algorithms. In F. Herrera and J.L. Verdegay, editors, *Genetic Algorithms and Soft Computing*, pages 279–305. Physica-Verlag, 1996.

92. J. Holland. *Adaptation in Natural and Artificial Systems*. University of Michigan Press, Ann Arbor, Michigan, 1975.

93. L. Holmström and P. Koistinen. Approximation capabilities of multi-layer feedforward networks. *IEEE Transactions on Neural Networks*, 3(1):24–38, 1992.

94. J. Horn, N. Nafpliotis, and D.E. Goldberg. A niched Pareto genetic algorithm for multi-objective optimization. In *Proceedings of First IEEE International Conference on Evolutionary Computation*, volume 1, pages 82–87, Piscataway, New Jersey, 1994. IEEE Press.

95. K. Hornik, M. Stinchcombe, and H. White. Multi-layer feedforward networks are universal approximators. *Neural Networks*, 2:359–366, 1989.

96. K. J. Hunt, D. Sbarbaro, R. Zbikowski, and P.J. Gawthorp. Neural networks for control systems - A survey. *Automatica*, 28:1083–1112, 1992.

97. K.J. Hunt and D. Sbarbaro. Neural networks for nonlinear internal model control. *Proceedings of IEE, Part D.*, 138:431–438, 1991.

98. M. Hüsken and B. Sendhoff. Evolutionary optimization for problem classes with Lamarckian inheritance. In *Seventh International Conference on Neural Information Processing*, Seoul, Korea, 2000.

99. M Hüsken and B. Sendhoff. Evolutionary optimization for problem classes with lamarckian inheritance. In *Proceedings of the 7th International Conference on Neural Information Processing*, 2000.

100. M.A. Huynen, P.F. Stadler, and W. Fontana. Smoothness within ruggedness: The role of neutrality in adaptation. *Proceedings of National Academic Sciences of USA*, 93:397–401, 1996.

101. C.-L. Hwang and A.S.M. Masud. *Multiple Objective Decision Making - Methods and Applications*. Springer, Berlin, 1979.

102. Y. Iiuni. A nonlinear regulator design in the presence of systems uncertainties. *IEEE Transactions on Neural Networks*, 2:410–417, 1991.

103. H. Ishibuchi and T. Murata. A multiobjective genetic local search algorithm and its applications to flowshop scheduling. *IEEE Transactions on Systems, Man and Cybernetics, Part C*, 28:392–403, 1998.

104. H. Ishibuchi, T. Nakashima, and T. Murada. Multi-objective optimization in linguistic rule extraction from numerical data. In E. Zitzler, K. Deb, and L. Thiele, editors, *Proceedings of First International Conference on Evolutionary Multi-criterion Optimization*, volume 1993 of *Lecture Notes in Computer Science*, pages 588–602, Berlin, 2001. Springer.

105. M. Ishikawa. Rule extraction by successive regularization. *Neural Networks*, 13(10):1171–1183, 2000.

106. A. Isidori. *Nonlinear Control Systems*. Springer, London, 1995.

107. M. Jamei, M. Mahfouf, and D.A. Linkens. Fuzzy control design for a bond graph model of a non-linear suspension system. In *International Conference on Bond Graph Modeling and Simulation*, pages 131–136, Phonix, AZ, 2001.

108. J.-S.R. Jang and C.-T. Sun. Functional equivalence between radial basis functions and fuzzy inference systems. *IEEE Transactions on Neural Networks*, 4:156–158, 1993.

109. J.-S.R. Jang and C.-T. Sun. Neuro-fuzzy modeling and control. *Proceedings of IEEE*, 83:378–405, 1995.

110. J.S.-R. Jang. Anfis: Adaptive-network-based fuzzy inference systems. *IEEE Transactions on Systems, Man, and Cybernetics*, 23(3):665–685, 1993.

111. C. Ji and S. Ma. Performance and efficiency: Recent advances in supervised learning. *Proceedings of IEEE*, 87(9):1519–1535, 1999.

112. F. Jimenetz, A. Gomez-Skarmeta, H. Roubos, and R. Babuska. Accurate, transparent, and compact fuzzy models for function approximation and dynamic modeling through multi-objective evolutionary optimization. In E. Zitzler, K. Deb, and L. Thiele, editors, *Proceedings of First International Conference on Evolutionary Multi-criterion Optimization*, volume 1993 of *Lecture Notes in Computer Science*, pages 653–667, Berlin, 2001. Springer.

113. Y. Jin. Decentralized adaptive fuzzy control for robot manipulators. *IEEE Transactions on Systems, Man, and Cybernetics*, 28(1):47–57, 1998.

114. Y. Jin. Fuzzy modeling of high-dimensional systems: Complexity reduction and interpretability improvement. *IEEE Transactions on Fuzzy Systems*, 8(2):212–221, 2000.

115. Y. Jin, J. Jiang, and J. Zhu. Neural network based fuzzy identification with application to modeling and control of complex systems. *IEEE Transactions on Systems, Man, and Cybernetics*, 25(6):990 997, 1995.

116. Y. Jin and J.P. Jiang. State estimation and adaptive control of multivariable systems via neural network and fuzzy logic. *AMSE Advances in Modeling and Analysis*, 43(2):18–22, 1994.

117. Y. Jin, T. Okabe, and B. Sendhoff. Adapting weighted aggregation for multi-objective evolution strategies. In *Proceedings of First International Conference on Evolutionary Multi-Criterion Optimization*, Leture Notes in Computer Science, pages 96–110, Zurich, March 2001. Springer.

118. Y. Jin, M. Olhofer, and B. Sendhoff. A framework for evolutionary optimization with approximate fitness functions. *IEEE Transactions on Evolutionary Computation*, 2000. Submitted.

119. Y. Jin, M. Olhofer, and B. Sendhoff. On evolutionary optimization with approximate fitness functions. In D. Whitley, D. Goldberg, E. Cantu-Paz, L. Spector, I Parmee, and H.-G. Beyer, editors, *Proceedings of the Genetic and Evolutionary Computation Conference*, pages 786–793, Las Vegas, Nevada, 2000. Morgan Kaufmann.

120. Y. Jin, M. Olhofer, and B. Sendhoff. Evolutionary dynamic weighted aggregation for multiobjective optimization: Why does it work and how? In *Genetic and Evolutionary Computation Conference*, pages 1042–1049, San Francisco, CA, 2001.

121. Y. Jin, M. Olhofer, and B. Sendhoff. Managing approximate models in evolutionary aerodynamic design optimization. In *IEEE Congress on Evolutionary Computation*, volume 1, pages 592–599, Seoul, South Korea, 2001.

122. Y. Jin, M. Olhofer, and B. Sendhoff. A framework for evolutionary optimization using approximate fitness functions. *IEEE Transactions on Evolutionary Computation*, 2002. in press.

123. Y. Jin and B. Sendhoff. Knowledge incorporation into neural networks from fuzzy rules. *Neural Processing Letters*, 10(3):231–242, 1999.

124. Y. Jin and W. von Seelen. Evaluating flexible fuzzy controllers via evolution strategies. *Fuzzy Sets and Systems*, 108:243–252, 1999.

125. Y. Jin, W. von Seelen, and B. Sendhoff. An approach to rule-based knowledge extraction. In *IEEE Proceedings of International Conference on Fuzzy Systems*, pages 1188–1193, Anchorage, Alaska, 1998. IEEE Press.

126. Y. Jin, W. von Seelen, and B. Sendhoff. On generating FC^3 fuzzy rule systems from data using evolution strategies. *IEEE Transactions on Systems, Man, and Cybernetics*, 29(6):829–845, 1999.

127. Y. Jin, W. von Seelen, and B. Sendhoff. Extracting interpretable fuzzy rules from RBF neural networks. Technical Report IR-INI 2000-02, Institut für Neuroinformatik, 2000.

128. Yaochu Jin. *Knowledge in Evolutionary and Learning Systems*. Shaker Verlag, Aachen, 2002.

129. Y.C. Jin and J.P. Jiang. Neural-network-based nonlinear feedback control. *Journal of Zhejiang University*, 27:158–166, 1993.

130. B. Johanson and R. Poli. GP-music: An interactive genetic programming system for music generation with automated fitness raters. In John R. Koza, Wolfgang Banzhaf, Kumar Chellapilla, Kalyanmoy Deb, Marco Dorigo, David B.

Fogel, Max H. Garzon, David E. Goldberg, Hitoshi Iba, and Rick Riolo, editors, *Proceedings of the Third Annual Conference on Genetic Programming*, pages 181–186, 1998.

131. J. Kacprzyk. Group decision making with a fuzzy linguistic majority. *Fuzzy Sets and Systems*, 18:105–118, 1986.

132. L.P. Kaelbling, M.L. Littman, and A.W. Moor. Reinforcement learning: A survey. *Journal of Artificial Intelligence Research*, 4:237–285, 1996.

133. A.B. Kahng and B. R. Moon. Toward more powerful recombinations. In L.J. Eshelman, editor, *Proceedings of International Conference on Genetic Algorithms*, pages 96–103, 1995.

134. N.N. Karnik, J.M. Mendel, and Q. Liang. Type-2 fuzzy logic systems. *IEEE Transactions on Fuzzy Systems*, 7(6):643–658, 1999.

135. C. Karr. Applying genetics to fuzzy logic. *IEEE AI Expert*, 6(2):26–33, 1992.

136. R.L Keeney and H. Raiffa. *Decisions with Multiple Objective: Preferences and Value Tradeoffs*. Cambridge University Press, Cambridge, UK, 1993.

137. H. Kikuchi, A. Otaka, and S. Nakanishi. Functional completeness of hierarchical fuzzy modeling. *Information Science*, 110:51–60, 1998.

138. H. Kitano. Designing neural networks using genetic algorithms with graph generation system. *Complex Systems*, 4:461–476, 1990.

139. J. D. Knowles and D. W. Corne. Approximating the nondominated front using the Pareto archived evolution strategies. *Evolutionary Computation*, 8(2):149–172, 2000.

140. T. Kohonen. *Self-organizing Maps*. Springer, Berlin, 1995.

141. Y.K. Kong and W.Q. Zhang. Fuzzy weather forecast model. *Chinese Journal of Fuzzy Mathematics*, 3:32–38, 1982.

142. B. Kosko. Optimal fuzzy rules cover extrema. *International Journal of Intelligent Systems*, 10(2):249–255, 1995.

143. B. Kosko. *Fuzzy Engineering*. Prentice Hall, 1997.

144. J. Koza. *Genetic Programming: On the Programming of Computers by Means of Natural Selection*. The MIT Press, Cambridge, MA, 1992.

145. J.R. Koza, F.R. Bennett III, D. Andre, and M.A. Keane. *Genetic programming III: Darwinian Invention and Problem Solving*. Morgan Kaufmann, 1999.

146. M. Kreutz, B. Sendhoff, and C. Igel. *EALib: A C++ class library for evolutionary algorithms*. Institut für Neuroinformatik, Ruhr-Universität Bochum, Bochum, Germany, 1.4 edition, March 1999.

147. A. Krogh and J.A. Hertz. A simple weight decay can improve generalization. In J.E. Moody, S.J. Hanson, and R.P. Lippmann, editors, *Advances in Neural Information Processing*, volume 4, pages 950–957. Morgan Kaufmann, San Mateo, 1992.

148. F. Kursawe. A variant of evolution strategies for vector optimization. In *Parallel Problem Solving from Nature*, pages 193–197. Springer, 1991.

149. Y.-J. Lai and C.-L. Hwang. *Fuzzy Multiple Objective Decision Making*. Springer, Heidelberg, 1994.

150. W. Langdon. *genetic Programming and Data Structures: Genetic Programming + Data Structures = Automatic Programming*. Kluwer, Boston, MA, 1998.

151. C. Langton, editor. *Artificial Life*. The MIT Press, Cambridge, Massachusetts, 1995.

152. S. Lawrence, C. L. Giles, and A.C. Tsoi. What size neural network gives optimal generalization? convergence properties of backpropagation. Technical report, Department of Computer Science, University of Maryland, 1996.

153. Y. LeCun. Generalization and network design strategies. Technical report, University of Toronto, 1989.

154. Y. LeCun, J.S. Denker, and S.A. Solla. Optimal brain damage. In D.S. Touretzky, editor, *Advances in Neural Information Processing*, volume 2, pages 598–605. Morgan Kaufmann, San Mateo, 1990.

155. C.S.G. Lee and M.J. Chung. An adaptive control strategy for mechanical manipulators. *IEEE Transactions on Automatic Control*, 29:837–840, 1984.

156. D.D. Leitch. *A new genetic algorithm for evolution of fuzzy systems*. PhD thesis, Department of Engineering Science, University Oxford, Oxford, UK, 1995.

157. K.-H. Liang, X. Yao, and C. Newton. Evolutionary search of approximated n-dimensional landscape. *International Journal of Knowledge-based Intelligent Engineering Systems*, 4(3):172–183, 2000.

158. C.-T. Lin and C.S.G. Lee. Neural-network-based fuzzy logic control and decision systems. *IEEE Transactions on Computers*, 40(12):1320–1336, 1991.

159. C.T. Lin and C.S.G. Lee. Real-time supervised structure/parameter learning for fuzzy neural networks. In *Proceedings of IEEE International Conference on Fuzzy Systems*, pages 1283–1290. IEEE Press, 1992.

160. Y. Liu and X. Yao. Ensemble learning via negative correlation. *Neural Networks*, 12(10):1399–1404, 1999.

161. E.N. Lorenz. Irregularity: A fundamental property of the atmosphere. *Tellus*, A(36):98, 1984.

162. R.D. Luce and P. Suppes. Preferences, utility and subjective probability. In R.D. Luce, editor, *Handbook of Mathematical Psychology*, volume III. Wiley, 1965.

163. S. Ma, C. Ji, and J. Farmer. An efficient EM-based training algorithm for feedforward neural networks. *Neural Networks*, 10:243–256, 1997.

164. T. Masui. Graphic object layout with interactive genetic algorithms. In *Proceedings of IEEE Workshop on Visual Languages*, pages 74–80, 1992.

165. K.E Mathias and L.D. Whitley. Initial performance comparisons for the delta coding algorithms. In *Proceedings of IEEE Conference on Evolutionary Computation*, volume 1, pages 433–438, 1994.

166. Z. Michalewicz. *Genetic Algorithms + Data Structure = Evolution Programs*. Springer, 3 edition, 1996.

167. M. Mizumoto and K. Tanaka. Some properties of fuzzy sets of type 2. *Information and Control*, 31:312–340, 1976.

168. J. Moody and C. Darken. Fast learning in networks of locally-tuned processing units. *Neural Computation*, 1:281–294, 1989.

169. P. Moscato and M.G. Norman. A 'memetic' approach for the Traveling Salesman Problem: Implemetation of a computational ecology for combinatorial optimization on message-passing systems. In M. Valero, E. Onate, M. Jane, J.L. Larriba, and B. Suarez, editors, *Parallel Computing and Transputer Applications*, pages 187–194, Amsterdam, Netherland, 1992. IOS Press.

170. H. Mülenbein and D. Schlierkamp-Voosen. Analysis of selection, mutation and recombination in genetic algorithms. In W. Banzhaf and F.H. Eeckman, editors, *Evolution and Biocomputation*, pages 142–168. Springer, 1995.

171. R. Myers and D. Montgomery. *Response Surface Methodology*. John Wiley & Sons, Inc., New York, 1995.

172. P.B. Nair and A.J. Keane. Combining approximation concepts with algorithm-based structural optimization procedures. In *Proceedings of 39th AIAA/ASMEASCE/AHS/ASC Structures, Structural Dynamics and Materials Conference*, pages 1741–1751, 1998.

173. K.P. Narendra and K. pathasarathy. Identification and control of dynamic systems using neural networks. *IEEE Transactions on Neural Networks*, 1:4–27, 1990.

174. D. Nauck, F. Klawoon, and R. Kruse. *Foundations of Neuro-Fuzzy Systems.* Wiley, Chichester, 1997.

175. A. Neumaier. Molecular modeling of proteins and mathematical prediction of protein structures. *SIAM Review*, 39(3):407–460, 1997.

176. S. Nolfi, J. Elman, and D. Parisi. Learning and evoltuion in neural networks. *Adaptive Behavior*, 3(1):5–28, 1994.

177. H. Nurmi. Approaches to collective decision making with fuzzy preference relations. *Fuzzy Sets and Systems*, 6:249–259, 1981.

178. T. Okabe, Y. Jin, and B. Sendhoff. On the dynamics of evolutionary multiobjective optimization. In *Proceedings of Genetic and Evolutionary Computation*, New York, 2002. in press.

179. S.A. Orlovsky. Decision making with a fuzzy preference relation. *Fuzzy Sets and Systems*, 1:155–167, 1986.

180. A. Ostermeier, A. Gawelczyk, and N. Hansen. A derandomized approach to self-adaptation of evolution strategies. *Evolutionary Computation*, 2(4):369–380, 1994.

181. I. C. Parmee, editor. *Evolutionary Design and Manufacturing.* Springer, London, 2000.

182. M. Powell. Radial basis functions for multi-variable interpolation: A review. In C. Mason and M.G. Cox, editors, *Algorithms for Approximation*, pages 143–167. Oxford University Press, Oxford, U.K., 1987.

183. L. Prechelt. Automatic early stopping using cross validation: Quantifying the criteria. *Neural Networks*, 11(4):761–767, 1998.

184. C.L. Ramsey, K.A.De Jong, J.J. Grefenstette, A.S. Wu, and D.S. Burke. Genome length as an evolutionary self-adaptation. In A.E Eiben, Th. Bäck, M. Schoenauer, and H.-P. Schwefel, editors, *Proceedings of 5th Conference on Parallel Problem Solving from Nature*, volume V, pages 345–353, Amsterdam, Netherlands, 1998. Springer.

185. K. Rasheed. Guided crossover: A new operator for genetic algorithm based optimization. In *Proceedings of IEEE Congress on Evolutionary Computation*, volume II, pages 1535–1541, 1999.

186. A. Ratle. Accelerating the convergence of evolutionary algorithms by fitness landscape approximation. In A. Eiben, Th. Bäck, M. Schoenauer, and H.-P. Schwefel, editors, *Parallel Problem Solving from Nature*, volume V, pages 87–96, 1998.

187. A. Ratle. Optimal sampling strategies for learning a fitness model. In *Proceedings of 1999 Congress on Evolutionary Computation*, volume 3, pages 2078–2085, Washington D.C., July 1999.

188. J. Redmond and G. Parker. Actuator placement based on reachable set optimization for expected disturbance. *Journal Optimization Theory and Applications*, 90(2):279–300, August 1996.

189. R. D. Reed and R. J. Marks II. *Neural Smithing: Supervised Learning in Feedforward Artificial Neural Networks.* The MIT Press, Cambrideg, MA, 1999.

190. R.D. Reed. Pruning algorithms - A survey. *IEEE Transactions on Neural Networks*, 4(5):740–744, 1993.

191. G. Rudolph. Convergence analysis of canonical genetic algorithms. *IEEE Transactions on Neural Networks*, 5(1):96–101, 1994.

192. M. Saerens and A. Soquet. A neural controller. In *Proceedings of International Conference on Neural Networks*, pages 211–214, 1989.

193. Y. Sano and H. Kita. Optimization of noisy fitness functions by means of genetic algorithms using history. In M. Schoenauer et al, editor, *Parallel Problem Solving from Nature*, volume 1917 of *Lecture Notes in Computer Science*. Springer, 2000.

194. T. Sasaki and M. Tokoro. Adaptation toward changing environments: Why Darwinism in nature? In P. Husbands and I. Harvey, editors, *Fourth European Conference on Artificial Life*, pages 145–153. MIT Press, 1997.

195. J.D. Schaffer. Multiple objective optimization with vector evaluated genetic algorithms. In *Proceedings of an International Conference on Genetic Algorithms and Their Applications*, pages 93–100, 1985.

196. R.E. Schapire. The strength of weak learnability. *Machine Learning*, 5(2):197–227, 1990.

197. N. Schraudolph and R. Belew. Dynamic parameter encoding for genetic algorithms. *machine Learning*, 9(1):9–21, 1992.

198. H.-P. Schwefel. *Numerical Optimization of Computer Models*. John Wiley, Chichester, 1981.

199. H.-P. Schwefel. *Evolution and Optimum Seeking*. Wiley, New York, 1995.

200. M. Sebag and M. Schoenauer. Controlling crossover through inductive learning. In Y. Davidor, H.-P. Schwefel, and R. Männer, editors, *Parallel Problem Solving from Nature*, volume III, pages 209–228. Springer, 1994.

201. B. Sendhoff. *Evolution of Structures: Optimization of Artificial Neural Structures for Information Processing*. PhD thesis, Institut für Neoroinformatik, Ruhr-Universität Bochum, Bochum, Germany, 1998.

202. B. Sendhoff, M. Kreutz, and W. von Seelen. A condition for the genotype-phenotype mapping: Causality. In Th. Bäck, editor, *Proceedings of the 7th International Conference on Genetic Algorithms*, pages 73–80. Morgan Kaufmann, 1997.

203. B. Sendhoff and M. Kreuz. Evolutionary optimization of the structure of neural networks using a recursive mapping as encoding. In G. Smith, N. Steele, and R. Albrecht, editors, *Proceedings of International Conference on Artificial Neural Nets and Genetic Algorithms*, pages 370–374. Springer, 1998.

204. R. Setiono and H. Liu. Understanding neural networks via rule extraction. In *Proceedings of the 14th International Joint Conference on Artificial Intelligence*, pages 480–485), Montreal, Canada, 1995.

205. M. Setnes, R. Babuska, U. Kaymak, and H.R. van Nauta Remke. Similarity measures in fuzzy rule base simplification. *IEEE Transactions on Systems, Man and Cybernetics*, 28(3):376–386, 1998.

206. J.S. Shamma and M. Athans. Analysis of gain scheduled control for nonlinear plants. *IEEE Transactions on Automatic Control*, 1990.

207. K. Sims. Artificial evolution for computer graphics. *Computer Graphics*, 25(4):319–328, 1991.

208. A.E. Smith and D.M. Tate. Genetic optimization using a penalty function. In S. Forrest, editor, *Proceedings of the 5th International Conference on Genetic Algorithms*, pages 499–505, San Mateo, 1993. Morgan Kaufmann.

209. W. Spears. Recombination parameters. In Th. Bäck, D.B. Fogel, and Z. Michalewicz, editors, *Evolutionary Computation*, volume 2, pages 152–169. Institut of Physics Publication, Bristol, 2000.

210. N. Srinivas and K. Deb. Multi-objective optimization using non-dominated sorting in genetic algorithms. *Evolutionary Computation*, 2(3):221–248, 1994.

211. S. Suddart and A. Holden. Symbolic neural systems and the use of hints for developing complex systems. *International Journal of Man-Machine Studies*, 35:291–311, 1991.

212. C.-T. Sun. Rule-base structure identification in an adaptive-network-based inference system. *IEEE Transactions on Fuzzy Systems*, 2(1):64–79, 1994.

213. P.D. Surry, N.J. Radcliffe, and I.D. Boyd. A multi-objective approach to constrained optimization of gas supply networks. In *AISB-95 Workshop on Evolutionary Computing*, 1995.

214. R.S. Sutton. Learning to predict by the methods of temporary differences. *Machine Learning*, 3(1):9–44, 1988.

215. S. Sutton and A. G. Barto. *Reinforcement Learning: An Introduction.* The MIT Press, Cambridge, MA, 1998.

216. G. Syswerda. Uniform crossover in genetic algorithms. In J.D. Schaffer, editor, *Proceedings of International Conference on Genetic Algorithms*, pages 2–9, Fairfax, VA, 1989.

217. G. Syswerda. A study of reproduction in generational and steady-state genetic algorithms. In G. Rawlins, editor, *Foundations of Genetic Algorithms*, pages 94–101, 1991.

218. H. Takagi. Interactive evolutionary computation. In *Proceedings of 5th International Conference on Soft Computing and Information*, pages 41–50, Iizuka, Japan, 1998. World Scientific.

219. T. Takagi and M. Sugeno. Fuzzy identification of systems and its applications to modeling and control. *IEEE Transactions on Systems, Man, and Cybernetics*, 15:116–132, 1985.

220. K. Tanaka. Stability analysis and design of fuzzy control systems. *Fuzzy Sets and Systems*, 45(2):135–156, 1992.

221. T. Tanino. Fuzzy preference relations in group decision making. In J. Kacprzyk, editor, *Multi-person Decision Making Using Fuzzy Sets and Possibility Theory*, pages 172–185. Springer, 1990.

222. T. Tanino, M. Tanaka, and C. Hijo. An interactive multi-criteria decision making method by using a genetic algorithm. In *Proceedings of 2nd International Conference on Systems Science and Systems Engineering*, pages 381–386, 1993.

223. K.M. Tao. A closer look at the radial basis function networks. In A. Singh, editor, *Conference Record of the 27th Asilomar Conference on Signals, Systems and Computers*, volume 1, pages 401–405, Los Alamitos, 1993. IEEE Computer Society Press.

224. G. Thimm and R. Fiesler. Pruning of neural networks. Technical Report 97-03, IDIAP, Lausanne, Switzerland, 1997.

225. S. Thrun and L. Pratt, editors. *Learning to Learn.* Kluwer Academic Publishers., 1997.

226. A. Tickle, R. Andrews, M. Golea, and J. Diederich. The truth will come to light: Directions and challenges in extracting the knowledge embedded within trained artificial neural networks. *IEEE Transactions on Neural Networks*, 9(6):1057–1068, 1998.

227. M. Tong and P.P. Bonissone. A linguistic approach to decision making with fuzzy sets. *IEEE Transactions on Systems, Man, and Cybernetics*, 10(1980):716–723, 1988.

228. E. Tunstel and M. Jamshidi. On genetic programming of fuzzy rule-based systems for intelligent control. *International Journal of Intelligent Automation and Soft Computing*, 2(3):273–284, 1996.

229. J. Valente de Oliveira. On the optimization of fuzzy systems using bio-inspired strategies. In *IEEE Proceedings of International Conference on Fuzzy Systems*, pages 1129–1134, Anchorage, Alaska, 1998. IEEE Press.

230. V.N. Vapnik. *The Nature of Statistical Learning Theory.* Springer, New York, 1995.

231. D. A. Van Veldhuizen and G. B. Lamont. Multi-objective evolutionary algorithms: Analyzing the state-of-art. *Evolutionary Computation*, 8(2):125–147, 2000.

232. G.P. Wagner, G. Booth, and H. Bagheri-Chaichian. Genetic measurement theory of epistatic effects. *Genetica*, 102/103:569–580, 1998.

233. J. Wakunda and A. Zell. A new selection scheme for steady-state evolution strategies. In *Proceedings of the Genetic and Evolutionary Computation Conference*, pages 794–801, Las Vegas, Nevada, July 2001. Morgan Kaufmann.

234. C. Wang, S. Venkatesh, and J.D. Judd. Optimal stopping and effective machine complexity in learning. In *Advances in Neural Information Processing Systems*, volume 6, pages 303–310. Morgan Kaufmann, San Mateo, CA, 1994.

235. L.-X. Wang and J. Mendel. Back-propagation fuzzy systems as nonlinear dynamic system identifiers. In *Proceedings of IEEE International Conference on Fuzzy Systems*, pages 1409–1416, San Diego, March 1992.

236. L.-X. Wang and J. Mendel. Generating fuzzy rules by learning from examples. *IEEE Transactions on Systems, Man, and Cybernetics*, 22(6):1414–1427, 1992.

237. C.J. Watkins and P. Dayan. Technical note: Q-learning. *Machine Learning*, 8:279–292, 1992.

238. http://www.lania.mx/~ccoello/EMOO/EMOObib.html.

239. A.S. Weigend, D.E. Rumelhard, and B.A. Huberman. Generalization by weight-elimination applied to currency exchange rate prediction. In R. Lippmann, J. Moody, and D. Touretzky, editors, *Advances in Neural Information Processing*, pages 875–882. Morgan Kaufmann, San Mateo, 1991.

240. P.J. Werbos. Backpropagation: Past and future. In *Proceedings of the IEEE International Conference on Neural Networks*, pages 343–353, San Diego, 1988. IEEE.

241. T. White and F. Oppacher. Adaptive crossover using automata. In Y. Davidor, H.-P. Schwefel, and R. Männer, editors, *Parallel Problem Solving from Nature*, volume III, pages 229–238. Springer, 1994.

242. D. Whitley, S. Gordon, and K. Mathias. Lamarckian evolution, the Baldwin effect and functional optimization. In Y. Davidor, H.-P. Schwefel, and R. Maenner, editors, *Parallel Problem Solving from Nature*, volume III, pages 6–15. Springer, 1994.

243. W. Wienholt. Improving a fuzzy inference system by means of evolution strategy. In B. Reusch, editor, *Fuzzy Logik*, pages 186–195. Springer, 1994.

244. L. Xu, S. Klasa, and A.L. Yuille. Recent advances on techniques static feedforward networks with supervised learning. *International Journal of Neural Systems*, 3(3):253–290, 1992.

245. L. Xu, A. Krzyzak, and E.Oja. Rival penalized competitive learning for clustering analysis, rbf net and curve detection. *IEEE Transactions on Neural Networks*, 4(4):636–649, 1993.

246. R.R. Yager. On ordered weighted averaging aggregation operators in multicriteria decision making. *IEEE Transactions on Systems, Man, and Cybernetics*, 18:183–190, 1988.

247. R.R. Yager and D.P. Feliv. Analysis of flexible structured fuzzy logic controllers. *IEEE Transactions on Systems, Man, and Cybernetics*, 24:1035–1043, 1994.

248. X. Yao. A review of evolutionary artificial neural networks. *International Journal of Intelligent Systems*, 8(4):539–567, 1993.

249. X. Yao and Y. Liu. A new evolutionary system for evolving artificial neural networks. *IEEE Transactions on Neural Networks*, 8(3):694–713, 1997.

250. J. Yen. Fuzzy logic - a modern perspective. *IEEE Transactions on Knowledge and Data Engineering*, 11(1), 1999.

251. J. Yen, L. Wang, and W. Gillespie. Improving the interpretability of tsk fuzzy models by combining global learning and local learning. *IEEE Transactions on Fuzzy Systems*, 6(4):530–537, 1998.

252. H. Ying. Sufficient conditions on uniform approximation of multivariate functions by general takagi-sugeno fuzzy systems with linear rule consequent. *IEEE Transactions on Systems, Man, and Cybernetics*, 28:515–520, 1998.

253. L.A. Zadeh. Fuzzy sets. *Information Control*, 8:338–353, 1965.

254. L.A. Zadeh. The concept of a linguistic variable and its application to approximate reasoning. *Information Processing*, 8:199–249, 1975.

255. L.A. Zadeh. The role of fuzzy logic in the management of uncertainty in expert systems. *Fuzzy Sets and Systems*, 11:199–227, 1983.

256. S.Q. Zhu. *Approximating analysis for nonlinear systems*. Higher Education Press, 1980.

257. E. Zitzler, K. Deb, and L. Thiele. Comparison of multi-objective evolution algorithms: empirical results. *Evolutionary Computation*, 8(2):173–195, 2000.

258. E. Zitzler and L. Thiele. Multi-objective evolutionary algorithms: A comparative case study and the strength Pareto approach. *IEEE Transactions on Evolutionary Computation*, 3(4):257–271, 1999.

Index

CPSIA information can be obtained at www.ICGtesting.com
Printed in the USA
LVOW100501280313

326427LV00003B/100/P

9 783790 825206